NUMERICAL CONTROL
FUNDAMENTALS

Jack Moorhead
Editor

Published by:
Computer and Automated Systems
Association of SME
Marketing Services Department
One SME Drive
P.O. Box 930
Dearborn, Michigan 48128

NUMERICAL CONTROL FUNDAMENTALS

Copyright, 1980 by the
Society of Manufacturing Engineers
Dearborn, Michigan 48128

First Edition

Library of Congress Catalog Card Number: 80-52723

International Standard Book Number 0-87263-057-9

Manufactured in the United States of America

SME wishes to express its acknowledgement and appreciation to the following publications for supplying the various articles reprinted within the contents of this book:

Cutting Tool Engineering
A Technifax Publication
P.O. Box 937
120 N. Hale Street
Wheaton, IL 60187

Machine and Tool Blue Book
Hitchcock Publishing Company
Hitchcock Building
Wheaton, IL 60187

Manufacturing Engineering
Society of Manufacturing Engineers
One SME Drive
P.O. Box 930
Dearborn, MI 48128

Modern Machine Shop
600 Main Street
Cincinnati, OH 45202

Production
Bramson Publishing Company
Box 101
Bloomfield Hills, MI 48013

Production Engineering
A Penton/IPC Publication
Penton Plaza
Cleveland, OH 44114

Tooling & Production
Huebner Publications, Inc.
5821 Harper Road
Solon, OH 44139

Grateful acknowledgement is also expressed to:

Weber N/C Systems
11601 West North Avenue
Milwaukee, WI 53226

PREFACE

About 200 years ago, steam replaced human muscle as a source of power in running industrial machinery. The change, known as the Industrial Revolution, brought to mankind a new world which included far less backbreaking work and increased productivity.

Many years later, a step which many scientists feel was the beginning of the second Industrial Revolution appeared on the manufacturing scene.

That step was Numerical Control.

Numerical Control improved machine control to the extent that non-productive efforts and wasted time were reduced. Before NC, machine operation depended on the skill of the operator. Quality was rarely consistent. NC allowed operators to ''lock in'' their best efforts, to be used again with the same predictable results.

Numerical Control, by definition, is the control of the actions of a machine by the insertion of coded commands. This coded input is usually in the form of numbers or letters. The system automatically interprets this data (usually with a small computer), and operates the machine.

The first numerically controlled machine appeared in 1952 at the Servo Mechanisms Laboratory of the Massachusetts Institute of Technology. This three-motion milling machine was a response to the need of the United States Air Force to consistently produce aircraft components with close accuracy.

In 1955, the Air Force began awarding contracts to build approximately 100 numerically controlled milling machines. The early computers used in those machines required a large amount of floor space. The computers were expensive, not easily programmable, and very susceptible to breakdown.

But as computers improved, so did NC machining. By 1960, NC machining appeared on the scene on a reasonably wide basis. As computing costs decreased and computer capability increased, NC uses boomed.

Today, over 9,000 manufacturing plants have installed their first numerically controlled machine. However, there are over 30,000 plants which could profitably use NC machines.

The purpose of this volume is to provide a solid foundation of facts for the engineers in those 30,000 plants. The book will help readers in their planning for the installation of NC. It is imperative that a successful NC application be proceeded by an extensive amount of planning.

The difficulty in arranging technical papers and articles into a logical succession of topics is that they have been prepared by people with a wide range of backgrounds, experiences, approaches, interests, etc. Thus, I suggest that the reader accept the specifics where the item deals with a common experience, and accept the general idea of what is being accomplished where the item deals with something unrelated to the reader's experience and applications.

Since its inception, NC has needed the type of person who has a strong desire to be a true professional. Hopefully, this volume will help start new people into the NC field, and advance others toward their own professionalism.

Chapter One of this volume presents an overview of Numerical Control, including material on building the NC teams and an introduction to programming.

Chapter Two, Part Programming, provides some hints to help avoid programming growing pains, and supplies an overview of the entire NC programming process.

As computer costs decreased, NC became more and more available to the small shop. Chapter Three, Small Shop Uses, discusses some of the special problems faced by small shops, and relates some of the ways to improve NC machine performance in a job shop.

One of the most neglected facets of NC—particularly by first-time users—is tooling. Over three dozen pages of Chapter Four, Tooling, have been dedicated to a discussion of this important topic.

Chapter Five discusses the NC approach to one of the foremost problems in America—productivity.

Finally, this volume presents some tips on reducing downtime. The final chapter, Maintenance, also outlines a preventative maintenance program.

I wish to thank the authors of the technical papers and journal articles which appear in this book. I also wish to acknowledge the publications who generously allowed us to use their material. They are: *Cutting Tool Engineering*, *Machine and Tool Blue Book*, *Manufacturing Engineering*, *Modern Machine Shop*, *Production*, *Production Engineering* and *Tooling & Production*. Finally, my thanks is also extended to Bob King of the SME Marketing Services Department for his efforts in producing this book.

Jack Moorhead
Editor

CASA

The Computer and Automated Systems Association of the Society of Manufacturing Engineers. . .CASA of SME. . .is an educational and scientific association for computer and automation systems professionals. CASA was founded in 1975 to provide for the comprehensive and integrated coverage of the fields of computers and automation in the advancement of manufacturing. CASA is the organizational "home" for engineers and managers concerned with computer and automated systems.

The Association is applications-oriented, and covers all phases of research, design, installation, operation and maintenance of the total manufacturing system within the plant facility. CASA activities are designed to do the following:

- Provide professionals with a single vehicle to bring together the many aspects of manufacturing, utilizing computer systems automation.
- Provide a liaison among industry, government and education to identify areas where technology development is needed.
- Encourage the development of the totally integrated manufacturing plant.

The application of computer/automated systems must always be timely and must assure cost-effective manufacturing and quality products.

CASA is an official association of the Society of Manufacturing Engineers. In joining CASA of SME, you become a partner with over 50,000 other manufacturing-oriented individuals in 35 countries around the world.

A member of CASA benefits from a constant output of data and services, including a discount on all CASA activities and SME books. CASA's educational programs, chapter membership meetings, publications, conferences and expositions have proven valuable by updating a member's knowledge and skills and by expanding his technical outlook in the integration of manufacturing systems.

Membership in CASA is a means for continuing education. . .a forum for technical dialogue. . .a direct channel for new ideas and concepts. . .an important extension of your professional stature.

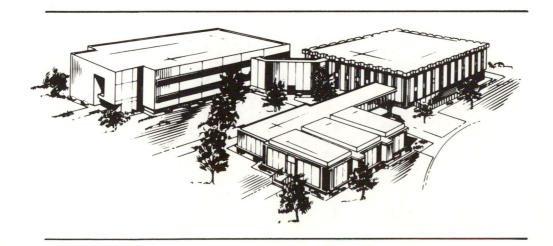

TABLE OF CONTENTS

CHAPTERS

1 INTRODUCTION

2 PART PROGRAMMING

3 SMALL SHOP USES

CHAPTER 1
INTRODUCTION

Reprinted from: Modern Machine Shop, December 1976

Building The NC Team Is A Management Job

By Ken Gettelman
Associate Editor
Modern Machine Shop

Karl Schultz
Manager—NC Programming Services
Cincinnati Milacron

An NC user may utilize many outside services in building his numerical control team, but one task he cannot farm out is the responsibility for organizing that team.

Mr. Schultz, as a person with considerable experience in observing both successful NC installations and a few disappointments, you often use the sports analogy. Why is this?

Aside from my own interest in sports, I cannot help but notice that the successful NC installation invariably has the same characteristics as a successful football or baseball team.

What are those characteristics?

A number of dedicated individuals, each capable at a certain task, working as a well coordinated team to achieve the goal of winning the game. Within a plant, it is making the NC installation function smoothly at a consistently high rate of productivity.

Is not the same thing true with manual machine operation?

There are basic fundamental differences and therein lies many of the misunderstandings and frustrations that have accompanied the introduction of NC in some plants.

Can you elaborate with your sports analogy?

Yes, we have all seen teams in baseball, football and other sports where there were individual stars who could do their jobs very well. An example is the baseball team with a single outstanding pitcher or home run hitter. He will win games and make a name for himself, but the team does not win championships. The winners are well-balanced and capable throughout and they function as a well-disciplined team with balance in all positions. Not only are they well balanced and disciplined, but they are well managed.

But how does this relate to machine tools?

The standard tool by its very concept is an individual rather than a team enterprise. Sure, the operator is trained and competent, but he has complete control over his machine. He has his own work methods, regardless of what it says on the process sheet, and he has the ability to adapt to unusual conditions during the machining process. If he is really good, he can do a good job. However, one of the hardest lessons to learn in numerical control is that NC in itself is a team effort. It is a totally different ball game than that of the conventional shop. To function to any degree of success, NC requires a good team effort, and it will surpass the performance of any individual or group with standard machines.

So what is the problem? Hasn't management of any business always put together a functioning team to achieve goals?

Good teamwork is essential to successful use of NC. It results from management commitment and management control.

This is the very nub of the matter. NC requires its own functioning team within the larger business team framework. The NC team will reach beyond the boundaries of the plant floor. Standard machine operation does not require that same kind of team organization. All too often shop management tends to treat NC machines as individual units in the hands of an operator rather than as an organized team effort of its own.

What positions are found on the NC team?

There are basically six: the selection of the proper NC machine, a good NC machine operator, the person or persons responsible for tooling, those who provide workpieces of expected size, those who develop and furnish a good part program, and the one who manages proper timing. Every person involved with NC will have a role with one or more of these playing positions.

What is the leadoff position?

The leadoff position is machine selection. Several people may have a hand in playing this important position. It is up to the team manager to see that all responsibilities of the position are properly executed.

What is the full scope of these responsibilities?

In addition to the actual act of purchasing, the NC team manager will have to secure satisfactory answers to a number of questions. For example:

- Is the machine proper for the present work load?
- Is it proper for the anticipated work load for the next five years?
- Does the installation site have proper floor support and adequate electrical supply?

-Are there proper provisions for lubricants, coolants, tooling and other accessory items?
-Are provisions made for maintenance and programming procedures and training?
-Has a preventative maintenance schedule been established and properly organized?

The leadoff position on the NC team is machine selection and several people may have a hand in playing this important role.

How about the second position of machine operation?

Here the NC manager should forget the idea that the bat boy can handle this position. The idea has been sold that anyone capable of pushing a button can fill this position. It simply isn't true. While the NC machine operator will not need the same skills as the operator of a conventional machine, it remains that the NC machine operator must have a sensitivity and knowledge of what is going on. When the unexpected happens he will then be able to save the day. Just as a spectacular catch by an outfielder can get a pitcher out of a jam, an astute NC machine operator can save the day in the event of broken tools, programming errors, unexpected differences in casting or forgings, and similar unexpected problems that may arise.

Your third position is tooling. How do you view it?

Tooling is a critical position on the NC team. The average conventional machine operator has his own favorite setups all laid out in his mind and no one else knows them nor is it essential to know. NC is a far more productive machining method but it is no longer an individual effort. It is a team game. Several people must interface with the machine. As players on a good baseball team have a communications system to avoid collisions and to decide who will back up whom, the NC programmer and operator must have a communications system for describing tooling and setups. This requires a tooling discipline with exact measurements. There must be a sufficient variety of tooling to accomplish the required jobs. Most of the turmoil surrounding the startup of a new NC machine will be avoided if there is good communication between the programmer and operator on tooling.

It is best to develop standard tooling packages that are known to both the operator and programmer. This requires a careful study of tooling required and a careful cataloging of them into a tooling system. Nothing is more embarassing to a baseball team than to have a fly ball drop between two players, each of whom thought the other was going to catch it. A tooling discipline is a communications system to avoid someone dropping the ball at the NC startup. Anyway one looks at it, teamwork is essential.

What about your fourth position, workpieces of an expected size?

In going back to the sports analogy, we find that most equipment is consistent. For example, each batter has his own favorite size and weight bat. The baseball is consistent. If anything changes, such as moving to a new ball park, or working with a different bat, the player wants to know. Ideally, nothing should change. But if it does, it should never come as a surprise.

The same holds true in workpieces for NC. Ideally, the workpieces should be those for which the program was written. If, however, there are size variations, it is important to tell the programmer and operator so that they can make the proper adjustments to avoid collisions or excessive tool wear or breakage.

Your fifth position is programming. How important is it?

Here is where good team management really shines and pays off. An NC machine is an impressive sight to watch. The beauty and motion of real machining productivity seem to come from the reel of one-inch perforated tape. The NC productivity, superiority and accuracy are obvious. What is not so obvious is the effort required to obtain that productivity increase.

The programmer is the person who controls the efficiency of the NC machine. A good programmer will provide economical, efficient and timely NC workpiece program tapes. A shop with a poor programmer will not be so fortunate. The programmer is the key to NC success. He must take the work of all the other players and combine it into a program for machining workpieces. He must then communicate the method to both the machine and the people who support the machine.

The NC machine operator must have a sensitivity and knowledge of what is going on. When the unexpected happens he will then be able to save the day.

Just as the baseball team manager looks for players with certain talents to bring balance and efficiency to the team, the NC manager will look for certain attributes in the programmer.

What are the most important things to look for in the programmer?

I think there are three: a machining background, a willingness and aggressiveness to do the job, and an analytical mind. All are equally important and all are necessary. (See MODERN MACHINE SHOP, June 1976, page 96, "Who Should Become Your NC Programmer?") Once the individual has been selected, do not hesitate to give him the proper training. Most NC vendors either provide the training or can direct your programmer to the proper training source. Do not make the mistake of sending a foreman or supervisor for NC training with the idea that he will train the in-plant programmer. This has proved, in most cases, to be a road to disaster. Something important always gets lost in the translation.

How about languages and computers for the programmer?

As soon as it becomes known that you've ordered an NC machine, salesmen from time-sharing services will come flocking to your door like player's representatives go running after new baseball prospects. There are a lot of NC computer tape preparation services available. Some are better than others, and some are more expensive than others. Often, if the NC user is considering computer assist, he is forced to make a decision based on little knowledge of what the language will really do, and how flexible it really is. The machine tool vendor can assist you; check with others in the area and have them program parts. It is usually safe to use a standard NC language such as APT. It covers all NC machine lines and is relatively easy to use. It can be run via time sharing or on in-house computer installations. The language that is chosen should allow for growth. The language and programmer must be able to meet the shop's workpiece load requirements.

If used properly, computer assist can greatly increase a programmer's productivity and accuracy. Use of such a service does not mean that the programmer does not need to have the attributes we have discussed nor a knowledge of manual programming; he must have these abilities plus the ability to work with the computer in its language.

Tooling is a critical position on the NC team. A tooling discipline is a communications system to avoid someone dropping the ball at the NC startup.

In your experience the programmer is a keystone position.

The programmer is the "field general" of the NC team. He must have cooperation from all of the other team members and, most of all, the authority to do his job.

Please define and explain the sixth position of "proper timing."

A 6-4-3 double play (shortstop to second base to first) is a beautiful thing to watch. The shortstop fields a ball which may bounce to him from any direction. He pulls it from his glove and times his throw to second just as the second baseman crosses the bag. The second baseman pivots and throws to first while jumping to avoid the cleats of the sliding runner. All of this must be done within a few seconds to get the runner at first and complete the double play. All three players must be in perfect harmony.

NC is the same in many ways. The availability of the machine and operator, the required tooling, the workpieces and an accurate NC program must all be available at the machine at the proper time. Any delay in any of these will cause downtime. Each of the players must have his job done, and each of the elements must be scheduled for arrival at the "on-deck circle" of the machine tool, with sufficient lead time, so that delays do not occur. In many new installations, delays are experienced because one element is missing. This can be reduced with a good qualified team and a good game plan.

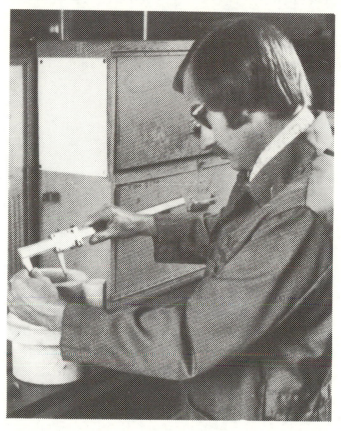

Workpieces must be of expected size to avoid trouble with a closely-tuned program; programmer and/or operator must be informed of any size variation before the job runs.

How does the manager fit into all of this?

Every team manager knows the importance of a good start. A good start inspires every member of the team. A lot of planning goes into preparing for the "season". Team players are evaluated and placed at their positions. Rookies are sent to rookie camp for individual training, and at spring training, the team effort is put together and team discipline is worked on. When the season starts, the team is ready for a series of encounters with the opposition.

The same is true of an NC startup. A good startup is important to obtain the desired results. A manager must be made responsible for the installation, and it is his job to select and send the programmers and maintenance people to "rookie" camp, and to insure that all of the facilities are available for the startup. He also studies and selects jobs to be run on the machine, and the tooling required. He coordinates the efforts of all the team members to insure that all is ready for the startup. With this "spring training", unexpected events at startup are minimized.

When the new NC machine hits the floor, it is your machine. It does not belong to the vendor or the bank. It is yours and it is your responsibility to make it work for you. The vendor is willing to provide you with many types of assistance. Use this assistance, but do not forget that you are the team manager, and the responsibility of the team and its success is yours. With good NC talent and good management, you should be high in the NC standings by the time mid-season arrives.

Introduction To NC Parts Programming

By R. J. Gaynor
and
U.I. VanBemden
Manufacturing Data Systems, Inc.

After carefully deciding on the right NC traveling wire machine tool to meet your requirements, careful consideration of the programming method should also be made. This will improve the scope and allow greater flexibility in supporting your NC facilities, both now and in the future. An introduction to the key factors that should be considered when evaluating the programming method will be made, beginning with an explanation of a Numerical Control program, and continuing with an in-depth look at the programming methods available, manual and computer assist, with examples of both.

INTRODUCTION

Choosing the right programming method is by no means an easy task, especially when one is unfamiliar with Numerical Control. Today's NC support market offers an abundance of different services, ranging from programming supplies to full service bureaus.

Also within this market are a number of companies dealing directly with computer assist parts programming, whose services range from time-sharing (renting computer time) to your own in-house system, which can vary from a microcomputer to a giant mainframe. So how does one make the right decision to meet one's needs now and in the future?

The best way to make this decision is careful consideration of the key factors that will directly affect your needs. Probably the most obvious factors deal with the lot sizes and complexity of parts your machine will produce. Other less apparent factors are lead time and machine load. Before going into depth on these and other key factors, perhaps it is best to refresh one's memory with the NC process.

The NC programming process begins with the part programmer studying the workpiece drawing. In effect he is much like a machinist, studying the part drawing before starting the machine and producing chips. However, he will not turn handwheels and operate controls as would the machinist. The programmer visualizes the machine motions necessary to machine the part. Once conceptualizing them, he will document them in a logical order. This information is then translated to the programming manuscript. He will do this whether the program itself is to be developed manually or with computer assist. The program is then checked by plotting out the coordinates from the programming manuscript. This may be done manually or via a computer assist NC graphic system.

Once the programmer is convinced that all his information is correct, the manuscript data is then converted to a medium (usually perforated tape) that can be input to the machine control unit (MCU). The programmer now has full responsibility for operating the machine tool effectively.

Now that you understand the programming process, it would be appropriate to define a Numerical Control program. By strict definition, Numerical Control is the operation of a machine tool by a series of coded instructions, which are comprised of numbers and other symbols. Webster's definition of a program is 'a logical sequence of operations to be performed.' Coded commands gathered together and logically organized so they will direct a machine tool in a specific task comprise an NC program. In other words it describes tool locations relative to the workpiece or the machine tool coordinate system. In addition to the description of the tool locations, other pertinent information must be given to the MCU. These additional instructions deal with rate and type of motion and any auxiliary operations the machine tool has.

The tool locations are relative to a coordinate system. This is usually the Cartesian coordinate system, also known as rectangular coordinates. (See Figure 1.) The concept fits machine tools

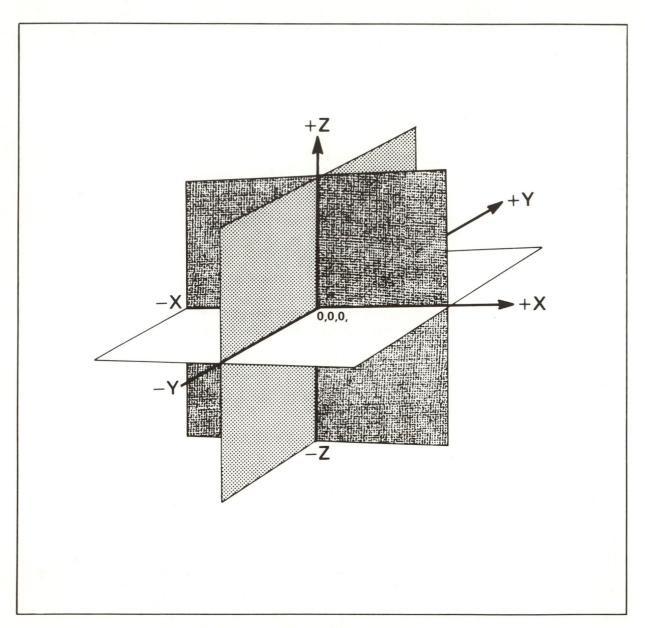

Figure 1: Coordinate System

perfectly. Machine tool construction is normally based on two or three perpendicular planes, with the common point of intersection being 'absolute zero.' The intersection of each pair of planes forms a line in space referred to as the X, Y or Z axis. An easy way to remember this system is the 'Right-hand Rule.' (See Figure 2.) For traveling wire machines we will only use two planes, since there is no up or down tool motion (Z movement). These two planes will form an X and Y coordinate system.

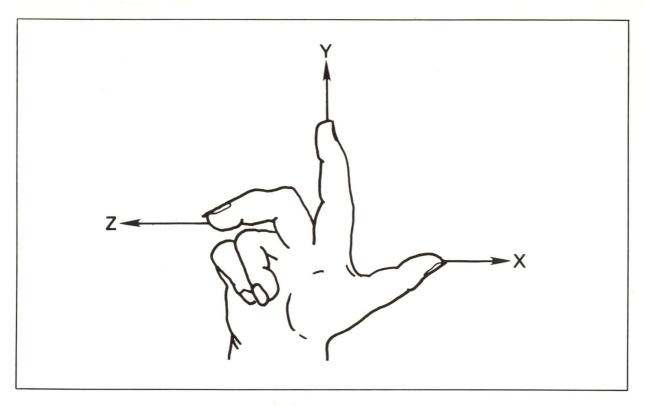

Figure 2: Right-hand Rule

The tool locations will be described with X and Y values relative to the absolute zero. The additional instructions dealing with type of motion (Preparatory Functions) and auxiliary operations (Miscellaneous Functions) will be described with G and M values respectively.

PROGRAMMING METHODS

There are basically two types of programming methods: manual and computer assist. The latter is broken down into numeral classifications based on language and type of processing.

MANUALLY

Manual part programming can be accomplished by an experienced individual familiar with the machine's operations. With today's large selection of NC machine/control configurations all requiring different input/output, it becomes more and more difficult for the manual parts programmer to meet the various programming requirements.

Another important factor to consider is the complexity of the part. This might determine whether manual programming is even feasible. With manual programming the programmer is forced to calculate the tool location for each point, a new point being required every time the tool path changes direction. A variety of considerations must also be taken into account by the programmer when adding the preparatory and miscellaneous functions to the program. Tool locations are calculated with geometry and trigonometry. Linear motion is illustrated in Fig. 3. Note that tool location 1, TL#1 is calculated by taking the absolute X-axis dimension 70 and adding the tool radius, giving an overall dimension of X76. The Y-axis location is calculated in the same manner. The second tool location, TL#2 is slightly more difficult to calculate. First you must calculate angle α in order to determine the θ angle.

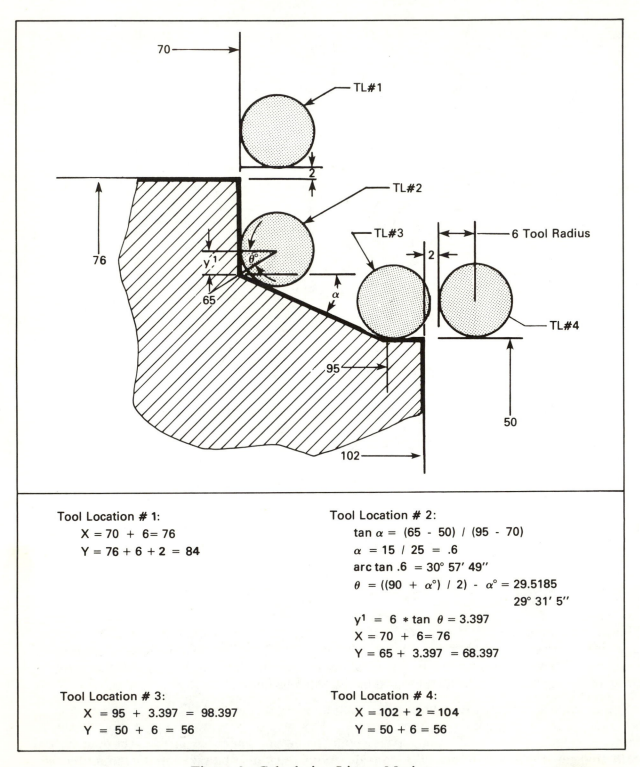

Tool Location # 1:
$$X = 70 + 6 = 76$$
$$Y = 76 + 6 + 2 = 84$$

Tool Location # 2:
$$\tan \alpha = (65 - 50) / (95 - 70)$$
$$\alpha = 15 / 25 = .6$$
$$\text{arc} \tan .6 = 30°\ 57'\ 49''$$
$$\theta = ((90 + \alpha°) / 2) - \alpha° = 29.5185$$
$$29°\ 31'\ 5''$$
$$y^1 = 6 * \tan \theta = 3.397$$
$$X = 70 + 6 = 76$$
$$Y = 65 + 3.397 = 68.397$$

Tool Location # 3:
$$X = 95 + 3.397 = 98.397$$
$$Y = 50 + 6 = 56$$

Tool Location # 4:
$$X = 102 + 2 = 104$$
$$Y = 50 + 6 = 56$$

Figure 3: Calculating Linear Motion

With the θ angle we are able to calculate distance y1, where y1 = tool radius x tan θ. Now y1 is added to the absolute Y-axis dimension of the two intersecting elements.

In determining the location of TL#3 the calculations are similar to the ones used for TL#2, whereas y1 is added to the absolute X-axis dimension of the two intersecting elements.

Calculating circular motion coordinates becomes somewhat more difficult, as you will note in Figure 4. This information is then transferred to the programming manuscript along with the preparatory and miscellaneous functions. Other pertinent machine control data is also added at

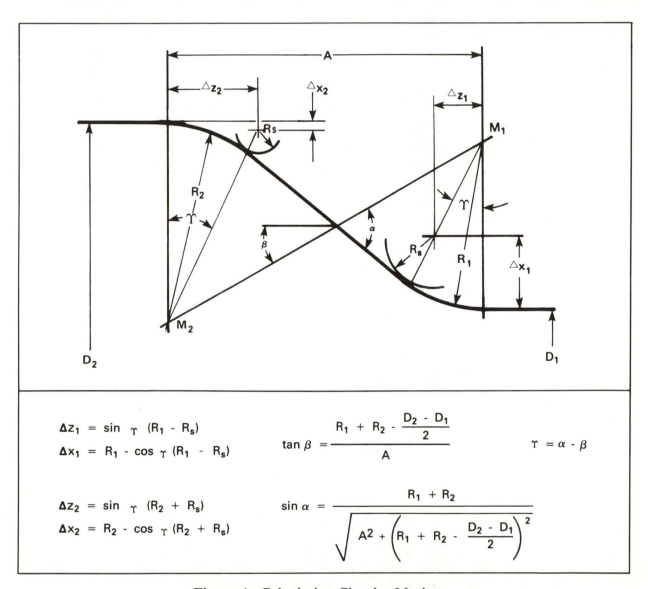

$$\Delta z_1 = \sin \Upsilon \ (R_1 - R_s)$$

$$\Delta x_1 = R_1 - \cos \Upsilon \ (R_1 - R_s)$$

$$\tan \beta = \frac{R_1 + R_2 - \dfrac{D_2 - D_1}{2}}{A}$$

$$\Upsilon = \alpha - \beta$$

$$\Delta z_2 = \sin \Upsilon \ (R_2 + R_s)$$

$$\Delta x_2 = R_2 - \cos \Upsilon \ (R_2 + R_s)$$

$$\sin \alpha = \frac{R_1 + R_2}{\sqrt{A^2 + \left(R_1 + R_2 - \dfrac{D_2 - D_1}{2}\right)^2}}$$

Figure 4: Calculating Circular Motion

this time. The information contained in the programming manuscript is then converted to codes on paper or mylar tape, which is read by the MCU. On programs containing a large amount of information, the opportunities for errors are greatly increased. The error possibility is then duplicated when converting the programming manuscript to machine code.

The ratio of errors inherited in part programs is not necessarily related to the program length; part complexity and machine tool control configuration are all contributing factors!

COMPUTER ASSISTED

Computer assist part programming is perhaps the only feasible method when contour machining. The programmer is able to describe part shapes, machine tool motions and machine functions in English-like commands. With computer assist programming, two steps are added to the NC programming process. The programming manuscript is now considered the source program, containing the input commands. This is converted to a medium, perforated tape, as was the programming manuscript, only now this tape is submitted to the computer for processing. **This is the first step.** The computer then produces a perforated tape containing similar information as that in the programming manuscript, which can be input to the machine control unit. **This is the second step.** The computer essentially makes calculations similar to the ones made by the manual programmer when preparing the programming manuscript. The computer makes certain geometric checks and solves geometric problems associated with part definition and tool path. In addition, certain computer systems might make some necessary logical decisions while automatically adding the necessary preparatory and miscellaneous functions in the proper sequence.

When using computer assist programming methods, parts are defined as a series of points, lines, circles, curves and planes. The tool path may be directed relative to the defined shape. The computer system compensates automatically for the tool size and shape. The input commands are translated by the software into the proper coding and tape format for the machine being programmed. This relieves the part programmer of these tedious and error prone activities.

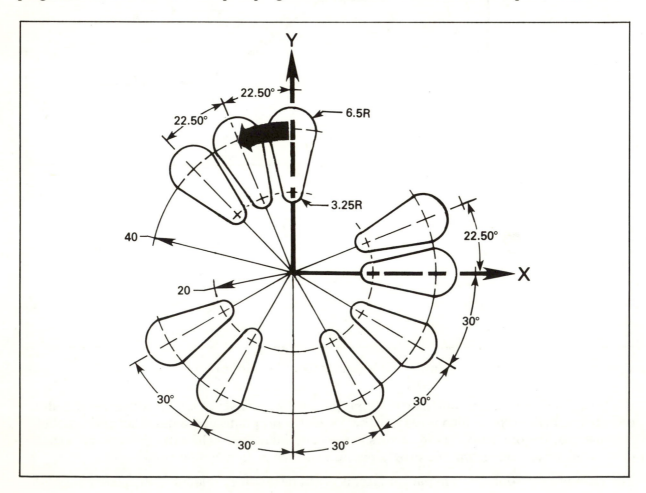

Figure 5: Sample Part

With a good and versatile computer system, the type of machine/control configuration makes very little difference since there is standard input which for all practical purposes remains basically the same.

The example (See Figure 5) shows a typical part suited for the traveling wire machines. The part actually consists of a very simple shape. The difficulty occurs when this shape is rotated around a common point, actually the center of the entire part, at different angles, nine times. By using the computer assist method of programming, the programmer is able to describe the motion once, at the 12 o'clock position, and then rotate that motion the remaining eight times, leaving the computer to calculate the rotated locations. (Figure 6.)

```
MACHIN,
IDENT, SHAPE #5
SETUP,LX,LY
DPT1,XB,40YB
DCIR1,PT1,6.5R
DCIR2,PT1,20YB,3.25R
DLN1,CIR1,CIR2,XS
DLN2,CIR1,CIR2,XL
MTCHG,.1TD
DVR1,0
DVR3,22.5
<1>MOVE,PT1,STOP
CONT,CIR1,90CCW,F(TANLN1)
  ; CIR2,S(TANLN1),F(TANLN2)
  ; CIR1,CCW,S(TANLN2),F89,STOP     $ GLUE STOP
  ; CIR1,CCW89,F90
MOVE,PT1,STOP
<2>DVR1,#1+#3
DO1/2,ROTXY-#1,2TMS
DVR3,30
DVR1,120
DO1/2,ROTXY-#1,2TMS
DVR1,210
DO1/2,3TMS,ROTXY-#1
DO1/2,ROTXY67.5
END
```

Figure 6: Computer Assist Part Program (COMPACT II)

KEY FACTORS - PARTS

Many of the key factors to consider when choosing the right programming method deal with the type of parts that will be produced on your NC machine. Probably the best approach is to analyze the parts themselves using the following guidelines:

1. Parts machined in small lots will create a greater demand for tapes since job runs are shorter.

2. High production parts will have longer runs and thus reduce the demand for tapes.

3. Parts with complex shapes will require more programming time which increases the error ratio while also increasing tape prove-out time.

4. One of a kind parts fall into the same category as the small lots, shorter job runs and greater demands for tapes.

5. Parts with similar shapes usually lend themselves quite well to the family-of-parts programming concept. Family-of-parts programs, also known as master programs, are ideal for the computer assist method. Changing the dimensional information from one part to another can be done quickly, easily and accurately via the computer's editing system. The program is then reprocessed producing a tape for the new part in a matter of minutes. To attempt such a task with manual programming usually results in a very time consuming assignment.

6. Parts with anticipated design changes could result in extensive modifications to the component with just minor changes to the machine tape.

7. The cost of human error on a very expensive part nearing completion is another factor to consider. Some computer assist systems provide comprehensive NC graphic capabilities that allow the programmer to plot the part geometry and cutter path prior to producing the machine tape, thus reducing the chance of program errors.

8. Parts that have complex configurations requiring close tolerances are more difficult to achieve with manual methods.

9. Parts that contain similar shapes that are mirrored, rotated or translated, present little advantage to the manual programmer, whereas these types of parts are ideal for computer programming systems which offer this type of capability.

KEY FACTORS - MACHINE TOOLS

The NC machine/machines also play an important part when choosing the programming method.

The following points should be considered:

1. The machine load has a direct bearing on the programming method. A machine that is operated for three shifts will require more tapes than one operated for just one shift.

2. Lead time for machine tapes is usually directly related to the programming method or system.

3. The number of machines and different machine/control combinations is also a factor. Manual programming methods require almost a one-to-one ratio programmer to machine/control combination, due to the variety of tape formats, numerical formats, miscellaneous and preparatory functions.

COMPUTER PROCESSING

When choosing a computer assist programming language or system, there are indeed some critical factors to consider. Probably the best place to start is with the type of computer processing and computer to be used. There are basically three types of computer processing available. A description of each of these is:

Batch Processing is when the programming manuscript, source program, is sent to a computer service organization (by mail or courier) or centralized EDP department. The programs are then gathered together and processed a batch at a time. Sending the programs back and forth for this type of processing is quite time consuming. This type of processing places an undue burden on the programmer because it is necessary for him to stay current with a number of programs that might be in process.

Remote Data Entry is another type of batch processing. Here information is transmitted by telephone lines to the service organization or centralized EDP department. Here again the information is gathered together and processed in batches. This is faster than the batch processing, but is still not perfect, for when errors occur the entire program has to be resubmitted for processing.

Interactive Time-sharing is the third type. Here the data is transmitted by telephone lines directly to the computer. Processing takes place immediately and the finished program is transmitted back to the customer's receiving device. With this type of system the programmer is able to complete a job from start to finish in one terminal session.

If you prefer to use or buy your own in-house computer, there are a number of systems available on the market today. Typically the smaller systems, microprocessor type, have less capabilities than the large mainframes. Although the minicomputer, which is slightly more expensive than the microprocessor, has all the capabilities the mainframes have.

When considering the computer assist programming language, there are over fifty from which to choose. Many of these are now obsolete and are not supported by any service companies. A large number are dedicated to a specific type of parts or to one specific machine tool. These usually have very limited capabilities and are impractical.

Probably the best type of language is the universal type. Here the same language can be used to program all the machine tools within one's facility. This also facilitates the expertise of the programmer, dealing with one language compared to two or three.

KEY FACTORS - LANGUAGES/SYSTEMS

Some of the key factors to consider before you select any language or computer system are:

1. Are arithmetic and trigonometric functions an integral part of the programming systems?
2. Is there an on-line NC graphic verification capability? Some systems have the capabilities of zooming in and blowing up a select area over 800 times scale.
3. Does the NC graphic capability include the drawing of the part geometry?
4. Does the system have bounded geometry? This will allow rotation, translation and mirroring of part shapes.
5. Is it a universal language, one which will support your other NC machines now and in the future?

6. Is there available software, links or postprocessors, to support your other machines?
7. Is there professional training available? If so, how long is it and what is the cost?
8. Is there on-site assistance available if necessary?
9. Is there advanced training available?
10. Is there a customer service available?
11. What is their experience in this field?
12. Is the system a proven product?

These are some of the prime factors to be considered when selecting the programming method or system. The final selection should be the result of comparative demonstrations of capabilities and time required using one or more actual parts. If several computer assist systems are to be considered, and they should be, insist on a start-to-finish demonstration in your shop, on your parts. Also request the names of shops in your area that do similar work or have similar machine tools, whom you may call for a recommendation.

Reprinted from: Modern Machine Shop, July 1977

Management Update:
Computers In Manufacturing

Developments are breaking fast and the options for the manufacturing manager are growing. Keeping abreast is a real challenge and an important one.

By KEN GETTELMAN, Associate Editor

It wasn't too many years ago that the manufacturing manager saw one application of the computer in the shop environment—a means of developing program tapes for his numerically controlled machine tools. Those days are gone. Today, the computer is used for everything from product design to final inspection. But the fact remains that the person responsible for machine tool operation is likely to place manufacturing computer applications in two distinct categories—NC and all other.

Indeed, this is the way industry as a whole tends to look at computers. The Numerical Control Society places most of its emphasis on machine tool operation while organizations such as CAM-I (Computer Aided Manufacturing-International) and the Society of Manufacturing Engineers take a much broader viewpoint. There are plenty of developments within the two frameworks. What was the state of the art one year ago has changed considerably within the past twelve months.

Numerical Control

It happened at the 1976 International Machine Tool Show in Chicago. CNC (computer numerical control) came into its own with a suddenness that surprised everyone. Control units with extensive internal computing capacity dominated the show floor. The inside of many control cabinets was mostly an empty air space. The extensive computing capacity resides in the radically new microprocessors. Small chips, only a fraction of an inch on a side, are able to carry the same capabilities as en-

tire computer assemblies of just a few years ago. The reduction in size and drastic lowering of cost have changed the computer hardware from a large expensive investment with moderate capability to a compact low-cost investment with copius capabilities.

Those playing the devil's advocate raise the question about the advantages of CNC. They question whether or not a CNC unit will actually run a machine tool any faster or differently than the original "hardwired" control unit concept that started the whole NC idea. The answer to running a machine tool faster has to be "no." Nor will CNC basically run a machine tool any differently. The methodology by which an NC machine tool operates and the speed with which it makes chips are more determined by the physical design of the machine itself than the numerical control unit directing it. But CNC has opened up whole new horizons for the NC machine tool and the method by which it is programmed and managed. An insight into the new potentials and the philosophical approaches to them were highlighted by two papers delivered at the Fourteenth Annual Meeting and Technical Conference of the Numerical Control Society in Pittsburgh. But first, a little background information.

With the exception of a few card type, magnetic tape, and wide tape numerical controls, the vast majority of all NC programs have been carried on the perforated one-inch-wide tape. Every coordinate position command and auxiliary machine command had to be completely stated in

a format that was compatible with the hardwired control and the machine tool it was directing. Each command was then punched into the tape. Depending upon the geometry of the workpiece, the tape could range anywhere from a few command blocks not more than two feet in length to a tape with hundreds of thousands of command blocks and thousands of feet in length for a complex aerospace part. Simple workpieces might be programmed manually but complex workpieces normally required the services of a computer to handle the vast amount of data that was processed.

It all sounds simple enough, but there were certain complexities and lack of a single standard approach. With the original hardwired control concept and design, an individual control unit could handle only incremental or absolute dimensioning statements. Originally, there were four different tape command formats. As time went on, a standard format was developed and even the hardwired controls could handle either incremental or absolute dimensioning statements. But how about the machine tool itself? Some were single spindle, some multiple spindle, some with automatic tool changers and so on. Also, different machine tools have different operating characteristics and dynamics. So in the final analysis, it was not only necessary to state where the cutting tool should go to machine a workpiece, but it was also necessary for the program tape to take into consideration all the individual characteristics of the control unit/machine tool combination. Thus, a program tape that would direct the machining of a workpiece on brand X machine tool with a brand Q control unit would not function on a brand Y machine tool with a brand P or even the same Q control unit.

Then along came improved electronics, which made possible the CNC concept; that is, a control unit's capability became a function of programmable software rather than the unit's physical hardware. But the expanded capabilities still did not

21

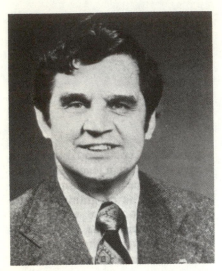

Kenneth L. Blair, Process Control Manager, Caterpillar Tractor Company

"The computer does far more than prepare program tapes for the NC machine tools."

change the basic concept of getting from a programmer's manuscript to a working tape. It was still necessary to go through processing and post processing whenever a computer was involved. Today, different options are in the wind.

The Three Steps

At present, most workpiece programs are still the result of a programmer studying a drawing and then deciding the basic methodology by which the machining will take place. If it is manual programming, he prepares a manuscript from which the program is punched by a person utilizing a tape preparation unit. If computer assist is employed, the programmer's statements are formatted for computer processing, which quite often still involves the standard perforated tape. Contrary to much erroneous belief, the initial computer output is not the finished program tape for machining the workpiece. The first step is nothing more than a definition of the path that the centerline of the cutter must follow to machine the workpiece. This is often known as the CL data— the general processing which defines the cutter path.

The CL data is a path definition and nothing more. Next comes one of the least understood aspects of numerical control. The CL data must be converted into a set of specific instructions that are appropriate for the particular machine tool/control unit combination that will actually machine the workpiece. It is here that the correct tape formatting is done. It is here that the particular configuration of the machine tool and its dynamics will be taken into consideration. This particular function is known as post processing—it follows the general processing.

When machine tool numerical controls were hardwired and rigid in capability, when tape preparation units were more mechanical than electronic, and when computers were large and expensive, the options were limited. Both processing and post processing had to be done on a large expensive computer. Some companies could afford their own. Most smaller companies had to time-share on a large system to distribute the cost among many users.

Many Options

With the advent of the microprocessor and the minicomputer, the single option of all processing and post processing on a large computer system has been eliminated. The case for moving post processing from the central computer to the CNC unit itself was made at the NC Society Annual Technical Conference by Lee Herndon of Vega Servo Controls, Inc., Troy, Michigan. It is Mr. Herndon's thesis that with flexible computing capacity in the CNC unit, it is now possible to put all the post processing instructions right in the control. Thus, only the CL data need be fed to the control and the control will then modify it before developing the proper machine operating instructions. The stated advantage would be the universality of a **CL data** for a workpiece. It could be taken to any CNC machine tool with the physical capability of handling the workpiece. No longer would program tapes be limited to use on a particular machine tool/control unit

Mr. George R. Armstrong, Vice President of Manufacturing, Caterpillar Tractor Co.

"The day of not understanding the computer as a powerful manufacturing resource is past."

combination.

There is a logical corollary that goes with the idea of putting the post processor software in the CNC machine control unit. Why not place both the processor language and the post processor in the CNC and introduce nothing but the programmer's source statements? Actually, this is now being done on a limited basis with certain types of machine tools— especially traveling-wire EDM units. A number of them have been built with minicomputers as an integral part of the CNC. Thus, the programmer's source statements are processed right at the machine tool with no need to utilize a separate computer source. If the methodology can become part of a two-axis machining unit, there is no reason why it cannot become a viable entity for three- and four- or even five-axis units.

Changes are rapidly taking place. Tape-punching systems that at one time were simple mechanical units have now become quasi computing terminals—thanks to the micro-processor. Data can be received, stored, changed, transmitted, and executed with the new breed of inexpensive terminal and programming unit combinations. Thus, data can be entered, verified, edited, and processed before it is ever committed to transmission or "dumping" onto tape or onto the transmission system to a remote computer.

Change, But No Change

Pointing out that the more things change, the more they are the same, Charles F. Raber of Manufacturing Data Systems, Inc., of Ann Arbor, Michigan, noted that while the methodology of handling NC software may be more flexible than ever with many new operating options on the way, the fact remains that software is still a key element of successful NC operation. He also pointed out that the job of developing and maintaining the ever-growing and changing body of NC software is still a highly specialized technique for the experts who make it available in an easy-to-utilize form for the NC machine tool user.

While any two NC users could discuss the merits of any given software handling approach, there is the new reality that the NC manager now has many more options open to him than he did yesterday and he will have even more tomorrow. He should be aware that new developments are coming and he should be open to any advantages they may offer.

CAD/CAM

While NC may have been the first application of the computer to the manufacturing environment, it certainly is not the last. Keeping track of all computer-aided design and computer-aided manufacturing developments is a significant task of its own.

CAM-I (Computer Aided Manufacturing-International), located in Arlington, Texas, is a non-profit organization made up of many national and international computer users who apply it to manufacturing—both NC and other-than-NC uses. CAM-I is an outgrowth of the old APT Long Range Program which was dedicated solely to improving the APT NC processor language. The documentation and administration of APT still rests with CAM-I but the organization's major thrust has gone to the non-machining areas. While some 20 or more

manufacturing computer software development projects are underway, a major effort has gone into the CAPP (computer-aided process planning) project. The first outline was demonstrated in early 1976 and some advancements were shown in Peoria, Illinois in early 1977. Attendees at the meeting saw the use of the manufacturing computer in both NC and other applications at the Caterpillar Tractor facilities in and around Peoria. In the East Peoria facility, complete DNC (direct numerical control) systems are used to control lines of NC machine tools so that a variety of workpieces going down the line will be shunted to the proper machine tool, which is controlled from the central computer.

The highlight of non-machining operations controlled by computer was seen in the new Caterpillar foundry located a few miles south of Peoria at Mapleton, Illinois. Computers control both the induction melting and holding furnaces so that electrical consumption is held at an even level and does not run into expensive peaks. This alone saves hundreds of thousands of dollars per year. Computers also control the sand mixing for the molds. The right blend of sand, binder, and sealer for each type of mold is completely computer controlled and dispensed from the storage bins to the computer controlled mold-forming lines. The pouring from the holding furnaces is also under computer control.

Thus, the foundry management knows just where it stands at any moment in the working day, week or month. The production computer keeps track of all schedules and output on a real-time basis and the information can be displayed on strategically located cathode ray tube readouts, which very closely resemble the common television tube.

Kenneth L. Blair, Process Control Manager at Caterpillar, told of the many computer projects within the corporation, which include computerized inspection, computer-controlled warehousing and computer-aided design in addition to the previously mentioned functions. These functions in themselves are both sig-

nificant and productive, but the full impact will be reached when the computer is utilized to coordinate and optimize all the various activities. This is the function of CAPP and Mr. Blair revealed how Caterpillar is committed to development and implementation of the CAPP program as a means of controlling manufacturing costs and keeping their facilities highly productive and efficient.

The whole theme was very well summed up by Mr. George R. Armstrong, Vice President of Manufacturing at Caterpillar, who gave an overview of the company's manufacturing operations around the world. He discussed the cost pressures, the energy problems, and the environmental concerns facing all manufacturing facilities today. He made note of the fact that the computer is a powerful tool in the manufacturing management's list of resources. He also noted that since it is such a significant resource, manufacturing management must make every effort to understand it and make use of it where applicable. The day of not understanding it is past. **MMS**

Reprinted from: Tooling & Production, July 1975

Helpful hints for NC programming, fixturing, etc.

Here's how to improve an improvement. Take a second look at your NC machining centers and gain a big boost in productivity.

by **Tony Rodway**
Production Engineer
Moog Inc.
Hydra-Point Div.
Buffalo, NY

AFTER you've automated, then what? Many shop managers achieve a dramatic improvement in productivity when they switch to NC. But, meanwhile, time passes, inflation gallops along, and before you know it, the boss says: "I know you saved us $50,000 last year, but what have you done lately?"

The way to head him off at the pass is to take a second look at each phase of your NC operation:

• Programming — improvements here cut machine cycle time.

• Tooling and fixturing—the aim is to reduce setup time, fixture costs.

• Optional equipment—you paid for it, be sure you reap the benefits.

• Support services—a sneaky one because the improvement started with the NC. However, focusing a spotlight on reduced materials handling, inventory costs, fixture maintenance and inspection procedures will show substantial savings that often go unrecognized.

Improvements in these areas have cut payout time by as much as 50 percent for many users of Hydra-Point NC machining centers. And they are relatively low-burden machines. So if the ideas discussed here make a difference on an inexpensive NC, the potential on the more expensive NC should now be proportionately higher.

Programming

On an NC machine with an automatic tool changer, a typical machining cycle includes both cutting and idling time. Cutting time can be shortened by optimizing speeds and feeds. Sounds simple, but a surprising number of shops fail to optimize the cutting function.

Finding the most efficient machining speeds paid off for Pelton and Crane Co., Charlotte, NC, makers of dental chairs. They balanced rapid chipmaking against excessive tool wear and cut per-piece cost in half, from over $50 per part to less than $25. This was on a part machined on a Moog Hydra-Point equipped with an automatic tool changer.

Analyzing programming can also lead to reduced machine cycle time by minimizing idling time—tool changing, spindle-speed changing and table positioning—but don't overlook the programming function itself. *Very often, when a programmer is pushed to get tapes out, he takes shortcuts that may save time for him, but could waste more expensive machining time.* Give the programmer sufficient time to work out the most efficient metal-cutting program. It reaps a reward every time you run the part.

Tool changing on a typical machining center takes 10 to 20 sec (chip-to-chip), but the table travels 5″ (12.7 cm) in only 1 sec. Better to move the table around to finish all the work to be done with one size tool than to switch tools just to complete one feature. And, since a random-access automatic tool changer requires a range of time to select a tool depending upon the proximity of the tools being exchanged, you can cut 4 to 6 sec out of a 10 to 20 sec change by loading tools so they are programmed in sequence. The changer moves only one station for the new tool instead of spinning half its circumference for one on the opposite side.

When the machine is waiting during a speed change, it could be positioning the table, accomplishing two nonmachining operations simultaneously. And be sure to program the farthest part of the operation first to allow time for the spindle speed to change; make the speed change during the long travel. Also, try to program the high-spindle-speed operations first—operations such as flycutting and milling—so that speeds can be gradually reduced as the part is being completed. Don't program the speeds to jump all over the lot, because each incremented change may take several seconds.

Tooling and fixturing

Good tooling and fixturing practices aim primarily to shorten setup time, but some, such as combined function tooling, preset tooling and multiple-part fixturing, also shorten machine cycle time.

One typical combination tool, for example, can spot-drill, drill and counterbore. It saves two tools and saves tool-changing time. But be sure to weigh savings in cycle time against cost of making, resharpening and replacing the tool—especially if you're cutting steel.

The preset tooling, of course, can be preset on a tool-setting block to required length before changeover, and thus save a lot of time between jobs.

Multipart fixturing can save both cycle time and setup time. If the workpiece is small, consider fixturing as many as possible on a pallet *off* the machine. Cycle time is shortened because each tool can be called for once to machine up to 36 parts, depending on how many can be mounted on the pallet. Set-up time is reduced to seconds because only the pallet must be mounted on the table while the machine is down. The individual part fixturing is done off the machine while it is working on another batch.

Larger parts are sometimes run two at a time. While the machine is working on one part at one end of the table, the operator fixtures a second part at the other end.

Subplates and special fixtures

Setup time can be substantially reduced if your NC machine has a subplate that contains dowel holes conveniently located at known positions in both X and Y directions. Once it is mounted on the machine table's T-slots, zeroed-in, and bolted down, locating reference points are fixed. Fixtures are quickly located on the proper dowel holes and bolted or clamped on.

In most applications where relatively long runs or repeat runs are common, special fixtures can be made for the workpiece. Once the fixture is made, setup is short relative to the length of the run.

Sometimes, however, runs are very short and consist of a large number of different parts. Here, special fixtures would price the job out of the shop.

Bob Carter of Marsland Engineering, Waterloo, Ontario, faced such a job. He had to tool up for **87 different parts, all of different shapes, and run 7, or multiples of 7, of each part** on his Hydra-Point 83-1000 MC machining center.

The completed assembly was a postal sorter containing over 4000 parts. Running seven at a time permitted delivery of finished units **at frequent intervals over the sev-**

Drilling holes in ends of six hanger arms. Automatic tool changer will then exchange drill for appropriate tap to complete the hole. Circle E4.

End milling the hook-shaped end of the hanger arms. They are still set up in the same fixture, but the indexer has rotated the parts 90 degrees.

eral years the jobs would run. This yielded significant savings in inventory costs and helped smooth out what could have been a horrendous assembly problem.

Bob solved his problem by separating all the parts into four families of shapes. He designed one simplified fixture for each shape: a double vise setup for squares and rectangles; a precision chuck for rounds; an angle plate for the third dimension of rectangles; and a V block for small-diameter rounds. Each fixture is permanently mounted on an adapter plate that in turn mounts on the machine's subplate. Once a fixture is on the table, all parts of that shape can be handled quickly.

Optional or special equipment

NC machining centers usually have a number of options that can be provided, but for extra cost. If you bought them, use them. Optional equipment for Moogs, for example, includes partial retract, peck drilling, fourth-axis indexers and automatic tape punches. With careful programming, the partial retract can save seconds each time a tool finishes one operation and

the table repositions for a second one.

Peck-drilling speeds deep-hole drilling while saving tool wear. And fourth-axis indexing, which makes it possible to machine a number of faces, reduces the number of setups and handling time.

The automatic tape punch, if available, not only permits editing of the tape on the machine, but can save zeroing-in time when castings vary slightly from part to part.

Support services

When you changed over from manual to NC machining centers, chances are you identified labor savings (both time and skill level required) and machine-time savings so that you came up with a per-piece cost substantially lower than with the old equipment. But you may have overlooked some hidden savings that could equal those you already identified.

These hidden savings are often overlooked because they are not so much NC related as they are "center" related. For example, if a job requires four different machines—drill press, milling machine, lathe and jig borer—and it can now be done on one NC machining center, you not only save on skilled

labor, setups and fixture costs, which you identified, but also on materials handling, in-process inventory and fixturing maintenance. These are the hidden savings.

Bill Thomas, production manager of Union Carbide's Florence, SC plant, calculated his hidden savings. He determined that 80 percent of his per-part production cost was in inventory. His 83-1000 MC's cut this cost by 75 percent because one machine does most of the work, thereby limiting queuing up of parts at each machine.

For the same reasons, materials handling was reduced 75 percent. The castings didn't have to be moved around as much. And workpiece-fixture costs were reduced because most parts could now be handled in a single setup on one machine. But a significant cost that is often overlooked is fixture maintenance, which Bill Thomas found to add up to 40 percent of the original cost of the fixture. So he reduced fixture *and* maintenance cost 90 percent.

Finally, inspection costs per part are lower. Fewer inspection steps are needed, and the NC assures a high level of repeatability, so inspection can be limited to random sampling of finished parts. ∎

CHAPTER 2

PART PROGRAMMING

Reprinted from: Production, June 1978

Computer-Aided Programming Relieves NC Growing Pains

What's new in Grand Haven, MI? Green Bay, WI? Hicksville, OH? According to Gardner-Denver Co., it's the successful introduction of computer-assisted NC parts programming techniques at these Pneutronics Div. locations

Only when the chips are flying is a numerically controlled (NC) machine tool paying for itself. Sitting idle, it's a budget buster. Gardner-Denver Co.'s Pneutronics Div. understands this well. It had NC machinery underutilized at three key locations.

The Pneutronics Div.'s investigation of machine tool underutilization centered on the Grand Haven, MI, plant where NC operations were representative of those at Green Bay, WI, and Hicksville, OH. This plant builds pneumatic tools and machinery for making solderless electrical connections.

It didn't take long for C. E. (Charley) Higdon, manager of special projects at Grand Haven, to zero in on the reasons behind machine tool underutilization. More and more work was being shifted from conventional machines to NC equipment; there were an insufficient number of part programmers; and point-to-point NC equipment was being upgraded to more difficult-to-program 3-axis and 4-axis machine tools.

NC Growing Pains. As Higdon correctly surmised, these events all were symptomatic of growing pains experienced by any company becoming involved with, and increasingly dependent upon, NC. But they were just that—symptoms. The reason for poor machine tool utilization was that equipment was sitting idle while NC part programs were being written and verified.

Manual Programming. Pneutronics Div. plants were programming NC equipment using an IBM-1130 computer, the limited ROMANCE language, and manual programming techniques. But as Higdon notes, manually programming even a relatively simple 2-axis NC machine tool can consume a lot of man and machine hours. Working from a blueprint, a programmer must calculate coordinates, type the program in machine language, punch the tape on a flexowriter, and dry run the tape on a NC machine as many times as necessary to insure its accuracy.

This approach to programming was fine for NC machine tools initially used by the Pneutronics Div. But it was obvious to management, says Higdon, that the division would increasingly have to depend on more sophisticated NC equipment to improve efficiency, reduce product cost, and remain competitive in the marketplace. However, a move to 3-axis and 4-axis NC lathes, machining centers, and jig bores represented a very large investment. And to insure the proper return on investment from such machinery, management felt that approximately 90 percent utilization of equipment was necessary. This presented some challenges. For example, approximately 5 to 6 new tapes, 35- to 50-ft long, would be required per day to load the machines as they were installed to

In the "old days," Gardner-Denver didn't worry about things like parts programming, tape verification, and data links. But today at its Pneutronics Div. plants, it is involved with all of this and more. It is this involvement with computer-aided parts programming, however, that permits the company to maintain approximately 90 percent utilization of its NC machine tools

prevent underutilization.

Time-Sharing. The Pneutronics Div. subscribed to computer-assisted NC parts programming to prevent a programming log jam and insure good machine utilization. In 1973, it entered into an agreement with Manufacturing Data Systems, Inc. (MDSI), Ann Arbor, MI, which tied the Grand Haven, Green Bay, and Hicksville plants into MDSI's central computer via remote time-sharing terminals.

What sold the division on MDSI, says Higdon, was its proprietary Compact II parts programming language. Compact II was compatible with all types and makes of NC equipment used by the three plants. Consequently, it was the only language programmers would be required to use. Learning the alphanumeric language was also quite simple.

With Compact II a programmer still works from a blueprint, but instead of manually writing a program in machine language, he prepares a description of the part, and the machining operations necessary to produce it, in the Compact II language, and enters it into an input-output terminal for transmission (via telephone lines) to MDSI. MDSI's computer generates the necessary machine tool codes and transmits it back to the remote terminal which instructs a high speed punch to prepare the machine control tape.

A strength of time-sharing is that the programmer is able to converse with the computer while the program is being generated. This allows much of the necessary editing to be performed before the tape is dry run on a machine. Each of the Pneutronics Div. plants also use a 3-dimensional plotter to graphically display the part being programmed in order to spot programming errors.

Time-sharing has proved an effective remedy for NC growing pains at the Pneutronics Div. plants. Higdon tells PRODUCTION that instead of making three or four tapes to obtain one good

Half a century ago, Pneutronics Div. plants were totally dependent upon the individual skills of machinists for the quality of parts. Today, engineering knowhow is stored in an in-house DEC computer and always retrievable

tape, only one and a half tapes is now needed for the typical program generated. This reduced tape turnaround time substantially cut setup time. The most noteworthy improvement, however, was the improvement in machine tool utilization. Higdon says time-sharing has reduced downtime attributed to tape errors by 30 to 40 percent.

In-house Computer. Computer-aided NC parts programming worked effectively for the Pneutronics Div. However, programming on a time-sharing basis became an economic burden about 2 years ago, says Higdon. That's when monthly user fees paid to MDSI began running $5000 to $6000 per month at Grand Haven, and between $12,000 and $14,000 per month for the three plants on the computer time-sharing network. Long-distance telephone charges for communicating with Ann Arbor also were becoming prohibitive. At Grand Haven, for example, they were approaching $800 a month.

These events, says Higdon, signalled the advent of more NC growing pains. "We had outgrown the time-sharing approach. We were now ready to bring parts programming to an in-house computer." To accomplish this, while re-

taining use of the Compact II programming language, the division became the first multilocation (3) user of MDSI's Action Central, an onsite minicomputer-based programming system.

Programming procedures under Action Central are essentially the same as those followed with time-sharing. Now, however, the program is generated by an onsite Digital Equipment Corp. PDP-1134 minicomputer (built to MDSI and Pneutronics Div. specifications) and equipped with the Compact II programming system.

Grand Haven has a number of Action Central machine tools (one link for each type of NC process used) with one to four machines connected to each link.

Higdon says that although Action Central has not directly reduced parts programming time, it has eliminated the problem of programmers having to wait to gain access to MDSI's computer. It also has lowered parts programming costs to within acceptable limits. And by having the minicomputer onsite, says Higdon, "we are able to make product and manufacturing engineering calculations with the equipment. This," he estimates, "saves Grand Haven about $1000 a month."

Brian D. Wakefield □

Master Family Programming—The Leading Edge In NC Utilization

By Robert W. Nichols
President
and
Larry Kanatzar
Vice President
NC-GT

MFP obsoletes single part programming, using Group Technology techniques. Definition of parametric CAD systems and how CAM uses parametric concept for N-C programming with MFP, to dramatically increase N-C programmers output. Computer program logic is illsutrated to explain MFP, also how management applies MFP successfully. Geometric Composite (GC) drawings from actual MFP show two axes turning development of GC and input sheets with pictures, as a follow-on from MFP part 1 MS77-992.

MASTER FAMILY PROGRAMMING

. MFP Obsoletes N-C Computer Programming For Single Parts

. Management & N-C Programmers Need Answers To NC-GT

   . What is it?

   . Why do you need it?

   . How do you apply it?

The use of computers for programming N-C equipment is accepted almost universally in the metalworking industry. However, startling new trends in Group Technology (GT) makes N-C programming for single parts obsolete. This fact is due to the rapid growth of computerized GT techniques, such as CAPPS for CAM, while in CAD there are numerous parametric software programs for family-of-parts, which can produce detailed drawings in seconds, even on stand-alone mini-computers. The first of these GT programs for CAD (using mini-computers) was PEP (Parametric Element Processor) developed by Computervision about four years ago. Then came GRIP (Graphics Interactive Programming Module) by United Computing Corp. Now, nearly all manufacturers of CAD/CAM systems have or are planning for this software capability.

Twenty years ago manufacturing proved money could be made by using a computer. This produced CAM. Then eight years ago CAD system and software developers were very creative; they included parametric software in nearly every system manufactured for mechanical/electronics designing and drafting, civil engineering, utility networks, map making, etc. Parametric as applied to GT is defined as a computer program for a family-of-parts, a program that uses one or many variable inputs to produce a complete drawing in CAD or a complete N-C tape/listing in CAM. Time savings using parametric programs may be 50 to 1 over conventional CAD increasing to 1000 to 1 on large complex drawings. The savings in CAM are similar using MFP compared to single part computer programming. The irony is that after 20 years CAM has not developed a parametric capability for any system that is on-line, fast, powerful and highly productive as CAD is with parametric software programs.

The new parametric concept for CAM is MFP. It will fit virtually every hardware/software system for N-C programming and is created for family-of-parts in true GT style.

The use of these types of GT parametric programs and MFP are part of the future that is currently emerging to increase N-C programmer productivity and N-C equipment utilization. This applies to batch production, and more so to R&D, maintenance and spare parts manufacturing.

Only companies who have applied GT will have eliminated the duplication of the many support functions by Manufacturing Engineering, and especially in the N-C computer programming process.

> Question: If GT is so great, do the following advanced
> production system/concepts use GT?
>
> MUM   (Methodology for Unmanned Manufacturing)
>
> ABMS   (Automatic Batch Manufacturing System)
>
> APAS   (Automatic Programmable Assembly System)
>
> I-CAM   (Integrated - Computer Aided Manufacturing)
>
> MFP   (Master Family Programming)
>
> The answer is yes, but

The definition of GT is:

1. One person manually (no computer) arranging in proper order a display (no coding and classification system) of all the information required to make cost savings and/or increased productivity programs possible. The time span could take one day or one year. This type of GT is "old hat" to manufacturing/ industrial engineers, tool designers, N-C programmers and others. This GT method is not systematic, no two approaches are ever the same, nor is it generally on-going. It's usually a "one shot" program, and is not intended for computer application.

2. The comprehensive GT systems apply coding and classification (C&C) techniques, that as a rule are programmed with a computer. Computerization makes possible GT programs that are beyond the ordinary ability of humans to manually retrieve and manipulate

such vast quantities of data. This has produced GT programs by the dozens to fit your product and organization.

There is a problem. These concepts and production systems do not use computerized GT systems to create, design and implement MUM, etc. The logic seems clear, these systems would be implemented in an even more professional manner using computerized GT. The advantages would be: increased management control, improved schedule, on-going, and upward capability for integrated CAD/CAM. Now CAPPS & I-CAM are a direct development of GT, as is MFP. However, MFP's are written using geometric composites that are created manually from large stacks of blue prints, but the best method to create a geometric composite is with a GT system using an on-line computer. This eliminates a great deal of time and manpower required to gather the blue prints and manually sort and group into families-of-parts.

We researched nearly 1,000 technical papers, magazines articles, and GT books. We talked to N-C programmers from coast to coast to find information about MFP type programming. There were various bits about the use of Macro, of which MFP is not. But nothing has been written detailing the required standardization of cutters and tooling, the use of logic flow charts, the elimination of set-up and pre-set documentation by the programmer, or the rare possibility of complete elimination of the N-C programmer. MFP is the master of the unfair advantage because it applies these types of GT discipline to the whole design and manufacturing system using a computer. It is a fact not widely recognized or understood that MFP will give you the leading edge by using GT techniques in N-C programming and equipment utilization.

The basics of MFP are:

1. Geometric composite of part families.

2. Standardizations of:

    A. Engineering design

    B. Manufacturing Engineering
        Methods and sequence
        Feed and speeds
        Cutters
        Tooling
        Equipment
        Programming
        Time standards

3. Program logic, and logic flow charts

4. Discipline - discipline - discipline, all of the above.

MFP is not a N-C computer program made up of a series of long complex macros. Instead it is the use of very simple logic statements that separate each standard program module. You notice we use the word standard again the reason is no trick programming methods are used, it generally looks like text book programming. This feature is especially suited to new program- mers, and makes program updating easy. The number of logic statements run from 30 to 500 or more. Since managers use logic daily as a work tool, why not apply logic to MFP suitability to increase your N-C productivity.

Logic statements used in many computer program languages (APT uniapt, adapt, compact II, and others) are composed of three basic elements and are executed in the following sequence by the computer.

1. IF
2. Subject (questioned by IF)
3. Logic or decision tree which checks values -, 0, + (of subject questioned by IF).

Let us apply this type of Computer Program Logic first to our- selves, then to MFP and, finally and most important, to the people who will make it work.

1. IF (we are "really sincere" about MFP) -, 0, +

-) I don't really want to look into MFP, but there is lots of money to be saved, also I could become a "hero" so better check a little further.

0) I cannot afford to be neutral, because the company has paid my way, and there may be chance to learn something new about applying GT to those damn expensive N-C machines.

+) I am sincere about MFP, so let's jump to the next logic statement.

2. IF (MFP) -, 0, +

-) No parts that look-a-like or fit a geometric composite, each part needs its own holding fixture and special non-standard cutting tools. Long production runs 10,000 parts or more. There is no pre-set cutting tool book or control of same. Do not have computer printout listing. Not using computer programming. STOP.

Go no further.

0) Afraid to make waves, not sincere, can't take the time, passive, doesn't care about N-C!

+) 1. Produces N-C tapes/listings in minutes.
   2. Eliminate tape and tool proofing.
   3. Reduces or eliminates N-C programming manpower.

4. Eliminate documentation (pre-set, set-up, operator instruction) by programmer.
5. Reduces N-C program back log.
6. Loads N-C equipment with proven tapes
7. Allows expansion of standard product sizes and materials without increased manpower (Manufacturing Engineer & N-C programmer).
8. Automatically justifies additional N-C equipment, if required.
9. Reduce scrap.
10. Standardizes:
    A. Process planning (sequence/method/equipment).

    B. Cutters all types, sizes, grades of material.

    C. Cutters location on machines.

    D. Holding fixtures.

    E. Feed/speed and depth of cuts (DOC) for each material and machine type/size.

    F. Programming.

    G. Time standards.

11. Synergistic (2 + 2 = 7) results (output) are greater than planned (input). Everything looks super but let's see if the next logic statement can stop this new NC-GT approach.

IF (MFP + Managment) -, 0, +

   -) Not successful, because management can not communicate to all levels of personnel the importance of the program to the company and how they are needed to help make it successful.

   0) Also unworkable if management is neutral. There has to be a positive management attitude, communications and a sincere belief in MFP.

   +) How management successfully implement MFP.

MANAGEMENT

Management recognizes that master family programming systems call for a high degree of cooperation between a number of groups and departments. Everyone in product design, process planning, tooling, N-C programming and the shop must realize how dependent each group is on the other. Everyone concerned must be informed of the plan, what the objective is, what part they play, and how it will affect them. Communications and cooperation are absolutely essential if master family programming is to work.

PRODUCT DESIGN ENGINEER

The design engineer works with the manufacturing engineer and
N-C programmer to develop the standard geometric shapes and
sizes for each family of parts. A composite drawing is used by
the programmer to write a master family program. This coordination between engineering and manufacturing establishes an effective method to keep costs down. MFP is a built-in monitor of
engineering changes or omissions from the product design standard.

MANUFACTURING ENGINEER

The manufacturing engineer assists and coordinates with the N-C
programmer in the development of the shapes, sizes and the geometric composite drawing. He establishes the operation sequence,
equipment type/size, tool number, gages, time standards, etc.,
that are a standard part of the manufacturing engineering support
functions which dovetail with the master family program. This
will eliminate a major part of the planning effort in the future.
You now have established master process planning sheets.

TOOL ENGINEER

The tool design engineer works with the programmer using the
composite drawing to create holding fixtures for all the sizes
and shapes of parts for one program for a single N-C machine.
This sometimes difficult task must be completed before the programmer starts making logic flow charts or programming.

CUTTING TOOL ENGINEER

The cutting tool engineer can be especially helpful in assist-
ing the N-C programmer with each cut, in the selection of each
cutter type, the grade of carbide, the correct feeds and speeds,
and depth of cut. This normally creates standardization of cutters.

N-C PROGRAMMER

The programmer has the most difficult task. He must be sure of
active support by management as he applies his special knowledge of computer program logic. He must have the correct information about the part geometry, planning, fixtures, cutters and
N-C machines. His task is to upgrade the cutting method to
achieve maximum feeds and speeds as well as spindle up time.
He is now a highly skilled "parts maker" due to the various inputs by communication with the shop and others.

N-C MACHINE OPERATOR

The N-C operator and foreman are very important to the success
of MFP. They are informed of the MFP concept and how they can
help. They work with the programmer to develop wording and

pictures on the computer listing for set-up and operator instructions, also review the method, tooling, operation sequence and feeds and speeds for possible improvements. This information and concurrence with N-C programmer and manufacturing engineer will give the confidence needed to eliminate all tape and tool proofing of MFP tapes in the future.

The N-C operaters enjoy working on MFP because it makes there work easier and more productive. The operators will find computer listings from MFP simple to understand, and easy to discuss with the programmer any suggestions or changes during debug and production start-up.

MFP are created using geometric composite (GC). We will illustrate a GC that is common to any valve manufacturing machine shop. It is a threaded shaft with various widths and locations of grooves and or thread reliefs. This is not a very complex GC but will quickly explain the concept. This GC and the MFP had produced over 250 tape/listings two years ago.

These GC reveal amazing facts about look-a-likes graphically. These pictures are the key to your MFP program and its success.

In actual experience we found you need to take a little extra time to draw your GC large, clear and use colors if possible to clarify each different family-of-parts.

The picture (GC) is now the work tool needed for coordination throughout the company. The GC should be developed by the N-C programmer. Then Manufacturing Engineering personnel need to develop standards with the machine tool process planner and tool cutter designer. The time standards should also be included. Then coordination with material control include purchasing on standard material sizes and finally we need product design engineers for concurance of the GC. After this a copy can be issued to Sales/Marketing, so they know what are the standard sizes, configurations, and material types. And sometime to everyone's great surprise no one really knew what was a standard. This standardizing is sometimes demanding to graduate engineers. They didn't have a course in the subject. So use salesmanship. You may have to remind them, you both work for the same company.

There are times when management uses these GC as a tool to establish standards. Its only a picture. The chief design engineers now clearly understand the GC, and especially those responsible in the design and drawing of all the future products, now the synergisms start, because the engineer knows we (Manufacturing engineering) have a GC, which he helped create, with a MFP that can produce machined parts in one day.

Only now the engineer discusses deviation from standard with the N-C programmer for possible options, such as new material, or a minor upgrade to the MFP, etc. The MFP with the GC has made the designer aware he may produce something that is not standard and will cost extra as well as impact schedules. Engineering is now <u>working with manufacturing</u> as a team to keep costs down and maintain standards. This doesn't mean you cannot make a non-standard part. It means the company is working with a system that gives management a graphic work tool with visibility and control over design and manufacturing. The reason is now a MFP can produce proven tapes at a very low cost and the N-C programming work force has been drastically reduced. Now when we get a non-standard part that cannot be programmed with a MFP the manufacturing engineering manager has to ask engineering and maybe sales for additional time (money) to program this special part. Surprisingly when these two managers get together many of these non-standards are drafting oversights, and so we are back to standardization by managers instead of designers. We know the designers will try harder next time.

There are several rules when developing a GC. Such as this one for turning on N-C lathe or chucker. There are of course different rules for 3, 4 & 5 axes type parts.

1. The maximum dia. (raw stock) for N-C machine, chuck or collet capacity.

2. The maximum length (including excess for facing) should include L/D ratio.

3. The number of material types.

4. The number of thread pitches (TPI) 20 - 18 - 16 - 14 - 12 - 11 - 10 - 9 - 8 - 7.

5. Do grooves fit standard groove tool or wider?

6. Are chamfers and fillet radii standard as constant, not variables.

7. Are all chan/radi's the same to break sharp edges as constant in the program to eliminate deburring.

8. The maximum bar length for bar type machines.

9. Is the part the same on each end for chucking end for end?

10. Are both ends of the shaft common to a single MFP?

There are three uses for the GC picture in N-C programming.

1. GC pictures with each variable, constants and all limiting parameters.

2. GC pictures with each symbol of N-C program geometry for programming.

3. GC pictures with variable to fit each separate configuration input sheet.

```
•••••••••••••••••••••VARIABLES•••••••••••••••••••••••••••••
50    OPT=0    $$OPTION •1,2,3,OR4
60    OD1=0            $$FIRST THD DIA.
70    OD2=0        $$MAJOR THD DIA.
60    OD3=0        $$THD RELIEF DIA. ZERO IF NO RELIEF
30    STKD=0       $$STOCK DIA.
90    D1=0         $$CHAMFER LENGTH ON END
10    D2=0     $$LENGTH ON FIRST DIA.
      D3=0     $$ ZERO IF OPTION 2
      D4=0     $$ ZERO IF OPTION 2 OR 3
      D5=0   $$ MIN PERF THD
      D6=0     $$ OPTION •4 SECOND THREAD RELIEF
60    TA1=0        $$CHAMFER ANGLE END
70    PIT=0        $$ THREAD PITCH
80    MAT=0    $$ 303 CRES=0, 316&416 CRES=-1,INCONEL X750=1
90    BMP=0    $$1=YES 0=NO (NO BUMPSTOP)
00    EXT=0        $$ STOCK EXTENSION FROM COLLET
10    SPD=0        $$ RPM FOR THREADING
```

OPTION-1

OPTION-2

OPTION-3

OPTION 4.

M F P
MASTER FAMILY PROGRAMMING

N-C MACHINE & MFP CAPACITY

- MAX STOCK DIA. 2-¼
     COLLET SIZES ⅜ TO 2¼ @ ½"
- MAX. EXTENSION OF STOCK FROM COLLET 7-¼"
- THREAD PITCHES PER INCH (TPI)
     7 - 8 - 9 - 10 - 12 - 14 - 16 - 20 - 24
- MINIMUM GROOVE WIDTH ³/₃₂" (.093)

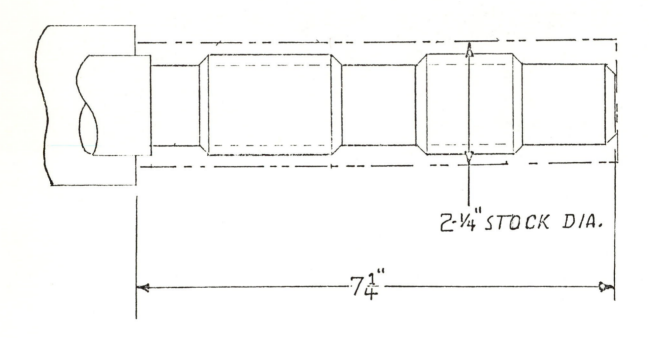

2-¼" STOCK DIA.

7¼"

PROGRAM GEOMETRY
THREADED SHAFT · VALVE STEM

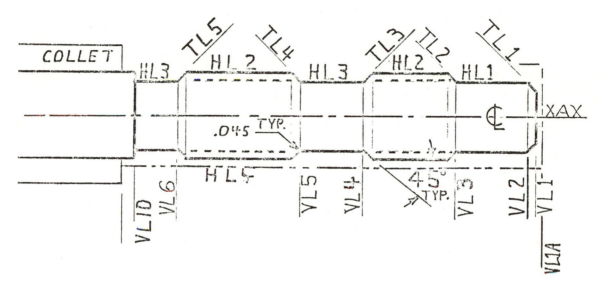

NOTE

ALL CONSTANTS ARE ADDED
FOR PROGRAMMER DOCUMENTATION

```
* * * * * * * * * * * * * * * * * * * *VARIABLES* * * * * * * * * * * * * * * * * * * * * * *
150     OPT=1      $$OPTION *1,2,3,OR4
160     OD1=_____        $$FIRST THD DIA.
170     OD2=_____        $$MAJOR THD DIA.
180     OD3=_____        $$THD RELIEF DIA. ZERO IF NO RELIEF
190     STKD=0       $$STOCK DIA.
200     D1=_____        $$CHAMFER LENGTH ON END
210     D2=_____$$LENGTH ON FIRST DIA.
220     D3=_____    $$ ZERO IF OPTION 2
230     D4=_____    $$ ZERO IF OPTION 2 OR 3
240     D5=_____  $$ MIN PERF THD
250     D6=0     $$ OPTION *4 SECOND THREAD RELIEF
260     TA1=_____       $$CHAMFER ANGLE END
270     PIT=_____       $$ THREAD PITCH
280     MAT=___ $$ 303 CRES=0, 316&416 CRES=-1,INCONEL X750=1
290     BMP=0    $$1=YES 0=NO (NO BUMPSTOP)
300     EXT=_____    $$ STOCK EXTENSION FROM COLLET
310     SPD=_____       $$ RPM FOR THREADING
```

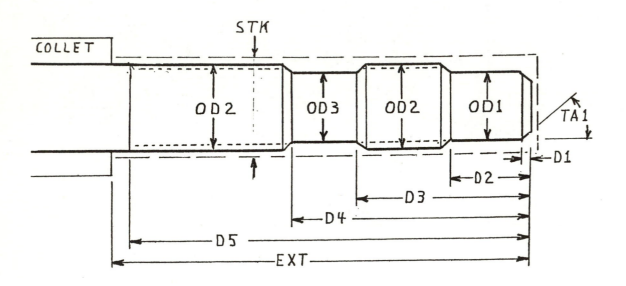

42

```
*  *  *  *  *  *  *  *  *  *  *  *  *  *  *  *VARIABLES*  *  *  *  *  *  *  *  *  *  *  *  *  *  *  *  *  *  *  *  *
150     OPT=2      $$OPTION *1,2,3,OR4
160     OD1=_____        $$FIRST THD DIA.
170     OD2=_____        $$MAJOR THD DIA.
180     OD3=0             $$THD RELIEF DIA. ZERO IF NO RELIEF
190     STKD=0      $$STOCK DIA.
200     D1=_____        $$CHAMFER LENGTH ON END
210     D2=_____     $$LENGTH ON FIRST DIA.
220     D3=0        $$ ZERO IF OPTION 2
230     D4=0        $$ ZERO IF OPTION 2 OR 3
240     D5=_____   $$ MIN PERF THD
250     D6=0        $$ OPTION *4 SECOND THREAD RELIEF
260     TA1=_____        $$CHAMFER ANGLE END
270     PIT=_____       $$ THREAD PITCH
280     MAT=___  $$ 303 CRES=0, 316&416 CRES=-1,INCONEL X......
290     BMP=0     $$1=YES 0=NO (NO BUMPSTOP)
300     EXT=_____  $$ STOCK EXTENSION FROM COLLET
310     SPD=_____     $$ RPM FOR THREADING
```

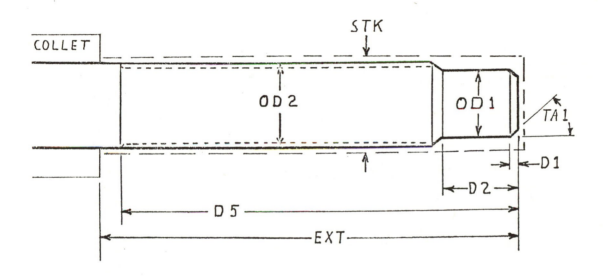

43

```
* * * * * * * * * * * * * * * * * *VARIABLES* * * * * * * * * * * * * * * * * * * * * * *
150    OPT=3    $$OPTION *1,2,3,OR4
160    OD1=_____        $$FIRST THD DIA.
170    OD2=_____        $$MAJOR THD DIA.
180    OD3=_____        $$THD RELIEF DIA. ZERO IF NO RELIEF
190    STKD=0       $$STOCK DIA.
200    D1=_____         $$CHAMFER LENGTH ON END
210    D2=_____   $$LENGTH ON FIRST DIA.
220    D3=_____    $$ ZERO IF OPTION 2
230    D4=0     $$ ZERO IF OPTION 2 OR 3
240    D5=0____ $$ MIN PERF THD
250    D6=0         $$ OPTION *4 SECOND THREAD RELIEF
260    TA1=____         $$CHAMFER ANGLE END
270    PIT=_____       $$ THREAD PITCH
280    MAT=___ $$ 303 CRES=0, 316&416 CRES=-1,INCONEL X750=1
290    BMP=0    $$1=YES 0-NO (NO BUMPSTOP)
300    EXT=_____   $$ STOCK EXTENSION FROM COLLET
310    SPD=_____       $$ RPM FOR THREADING
```

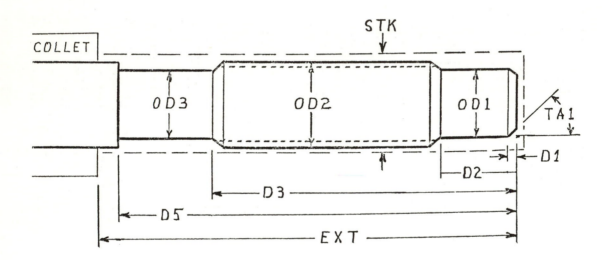

44

```
*****************VARIABLES*****************************
150    OPT=4    $$OPTION *1,2,3,OR4
160    OD1=1.375         $$FIRST THD DIA.
170    OD2=1.492         $$MAJOR THD DIA.
180    OD3=1.405         $$THD RELIEF DIA. ZERO IF NO RELIEF
190    STKD=1.75   $$STOCK DIA.
200    D1=.125          $$CHAMFER LENGTH ON END
210    D2=.39   $$LENGTH ON FIRST DIA.
220    D3=1.75     $$ ZERO IF OPTION 2
230    D4=2.125    $$ ZERO IF OPTION 2 OR 3
240    D5=3.87$$ MIN PERF THD
250    D6=.375    $$ OPTION *4 SECOND THREAD RELIEF
260    TA1=45           $$CHAMFER ANGLE END
270    PIT=12           $$ THREAD PITCH
280    MAT= 1   $$ 303 CRES=0, 316&416 CRES=-1,INCONEL X750=1
290    BMP=1    $$1=YES 0=NO (NO BUMPSTOP)
300    EXT=4.125   $$ STOCK EXTENSION FROM COLLET
310    SPD=450     $$ RPM FOR THREADING
```

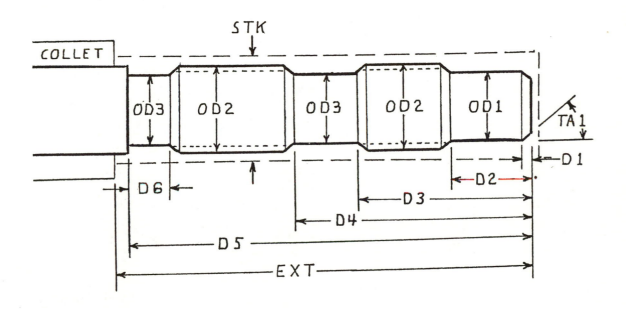

45

Prompt NC Language Offers Simplicity and Speed

By Albin Meske
Weber NC Systems

Since the original APT language for the programming of NC machines
was comissioned by the US Government 28 years ago, there are more than
20 major languages in common use. Each of them is either a version of
APT itself or else it bears strong evidence of having been developed
by a person who was familiar with APT.

PROMPT has been developed entirely outside of the APT family. It
bears no resemblance to the latter. The result is a startling difference in learning speed for this language, as well as high programming
speed.

This paper discusses some of the differences between PROMPT and its
counterpart languages and, in particular, explains the manner by which
geometry descriptions are made in contour machining.

INTRODUCTION

Although APT is an elegant and powerful means of generating NC tape,
it is also very complex. Although there are a number of simplified
NC languages available, the time required to gain proficiency in most
of them requires six months or more, particularly for the more powerful
languages.

From a managerial point of view, this has some unfortunate consequences.
A typical manager cannot afford to take a half year out of his life in
order to become conversant enough with a given language so as to effectively make intelligent decisions with regard to its choice. For many
companies, the initial choice of an NC language must be made in an
atmosphere of ignorance.

A further result of the above is that the NC part programmer experiences
a freedom from effective supervision by management which is accompanied
in equal measure by distrust and suspicion on the part of the latter.

The above factors were instrumental in leading to the development of
the PROMPT system of NC tape preparation.

DEVELOPMENT OF PROMPT

PROMPT was developed by John Weber who is the President and Owner of a
tool and die company specializing in plastic and die cast molds. He could
not afford the apprenticeship costs referred to above with regard to
learning the NC language in use in his shop. Being in a highly competitive business, he was uncomfortable in the knowledge that while NC
was playing an increasingly important role in his manufacturing operations, he did not have a way to evaluate the economic aspect of his

programming operations.

Tool shops pose the greatest test of programming performance because programming costs are proportionally greatest when the program must be amortized against a single machine cut.   In these circumstances the programming cost is often the major part of the total manufacturing cost. Mr. Weber was rightly concerned about the part that these costs played in determining his competitiveness.

Of equal importance was the exposure which he faced due to the fact that he was highly dependent upon the knowledge and skill of a man who could not be readily replaced if something happened to him. These fears were realized when his programmer quit to take another job.

For defensive reasons it was deemed expedient to send two men to the programming school being offered by the time sharing company. One of these men was the General Manager of the company, thus compounding the costs of additional training.

Perhaps the most significant single factor in the development of PROMPT was Mr. Weber's decision to ignore the current state of the art during the entire process of language development. Not only did he not survey the available languages, but he also steadfastly refused all offers of help and advice from his own trained programmers. He recognized that a knowledge of existing methods and solutions would color his own judgement and affect the resultant design.

The result of these efforts is PROMPT. There is virtually no similarity between it and the approach taken by existing languages.

INPUT FORMAT

The decision to develop an NC language coincided with the availability of the powerful and inexpensive desk top computer. PROMPT was developed on a Hewlett-Packard 9831A computer. In addition, a number of convenient programming features built into this instrument, it has an alphanumeric display window. The latter makes it possible to prompt each response, thus eliminating many of the traditional problems of language.

PROMPT can be said to have no language. Each response is elicited by a question appearing on the display. The question becomes part of the answer. Improper responses will be rejected. One cannot make a mistake due to language.

An example of the above is: TOOL CHANGE (Y OR N). An answer other than a Y or N will result in an audible beep and the question will again be repeated on the display.

This feature also makes high programming speed possible. Because the question is part of the answer, the latter can be kept to an absolute minimum. Except for numeric data input, responses are generally limited to a single character in which there is never a question as to the form which the response should take.

It is not even necessary to use the X, Y or Z keys in PROMPT programming. The program sequence is such that X, Y and Z coordinates and values are always input in response to a question. Thus, the programmer is spared the necessity of these additional key strokes.

CONTOUR DESCRIPTION IN PROMPT

One of the more radical innovations of PROMPT is its method of geometry description in contour machining.

PROMPT relies on eight simple figures - called geometric composites - by which part geometry is described.

The rules by which these composites are chosen and described are very simple. The author's eleven year old daughter learned their use and application in one evening. Customers usually learn the PROMPT contouring system in 2½ to 3½ hours. Many of them generate programs on the same day that the system was learned.

Figures 1, 2, and 3 show three PROMPT composites. Figure 4 illustrates a shape which will be described using the PROMPT system.

Starting at the point where the machine cut is to be initiated, a given shape is described in terms of PROMPT composities. Each is chosen and specified in the same order that the cutter would have encountered the corresponding shape on the workpiece. Each individual composite is input (on a prompted basis) to the computer in the order in which its elements, i,e, circles and line segments, would have been encountered by the cutting tool.

The above process is somewhat like the game of dominos. If a given tile is laid down and it ends in the number 6, the next tile must also begin with a number 6. In the same manner, if a particular composite ends in a .75R circle, then that same .75R circle must also represent the starting point of the subsequent composite.

The versatility of these PROMPT composites is further enhanced by the ability of the computer program to accept a zero radius in place of a finite value. Thus, in the case of the Type 2 composite shown in Fig. 1 Either of the circles shown in the figure can be input with a radius of 0. It follows that, in addition to representing two circles connected by a straight line which is tangent to both, this same composite will also represent a straight line which runs from a point to a tangent circle or vice versa - depending upon the order in which the data is input.

Whenever a geometric composite has been identified by inputting its Type

No., the computer will ask the X and Y coordinates of the ending point of the composite with respect to its beginning. These datums are always the centers of the circles. It will also request the size of each radius in the composite. Each will be input in the sequence of appearance as mentioned above. The computer will also ask the rotational sense of each circle (clockwise or counterclockwise) in order to properly locate the circle center.

In Figure 5, the inputs needed to define the shape illustrated in Fig. 4 are listed as they would appear in the computer printout. X and Y parity values are automatically calculated by the program to identify errors in X and Y inputs before any further steps are made in the program.

A complete description of PROMPT is beyond the capabilities of this paper. However, some of its more important attributes are listed:

> Full plotting capability, including offset cutter paths.
> Automatic generation of startup and ending codes.
> Offset calculation
> Rapid editing features permitting family of parts programs.
> Translation and rotation of shapes.

POINT TO POINT DESCRIPTION IN PROMPT

The PTP program in PROMPT offers equally innovative solutions to the job of describing part geometry and of generating machining cycle sequences.

The prompting feature of the program makes it possible for the NC programmer to choose and describe each machining cycle with a control of the detailed steps which is normally not possible or practicable with batch type languages.

In addition to plotting each coordinate center to be used in subsequent milling, drilling, boring and tapping operations, the plotter also draws in the tool motions and inscribes circles as they are specified by the programmer in the cycle description portion of the programming operation. Multicolor plotting distinguishes between milling, drilling boring and tapping operations.

In conventional NC languages, the computer input required to describe a PTP machining cycle will often exceed the input required to generate the same tape manually. This is not the case with PROMPT.

In customer trials, a PROMPT system will often result in the generation of a plotted, verified tape ready for machining long before the data for manual or computer assisted input has even been written down.

CUSTOMER EXPERIENCES WITH PROMPT

PROMPT programs are currently in use on machining centers, drilling machines, lathes, punch presses, flame and plasma arc cutters and wire EDM machines and jig grinders.

All customers report that PROMPT is substantially faster than the computer assisted or time shared language which they had previously used.

Of even more importance is the on PTP machining operations, PROMPT exceeds their former speeds for manual preparation of NC tapes. One reasons for this is that the prompted format greatly reduces the amount of input required.

SUMMARY

PROMPT is a revolutionary approach to the preparation of NC tapes. At present, PROMPT is limited to 2 axis contour generation with 3rd axis calculation of motions and feed rates. Although it is not as elegant as existing languages, it very obviously bears the imprint of having been conceived and designed by a person whose primary concern was that he would have to sign the checks for NC programming cost.

As such, PROMPT has met its original design objectives very well. Its simplicity and speed of learning will be of obvious importance to the harried executive who is often held hostage by his inability to take the time and effort required to learn the black art which traditional NC programming languages are regarded as being.

Although PROMPT is being used in both large and small sized companies, it is of particular help to the tool shop and the short run machinist. His entry into and participation of NC manufacturing is severely limited by both learning costs as well as programming costs.

It is expected that PROMPT will play a large role in promoting the growth of NC in this area.

Type No.	X	Y	R1	R2	R3	Rot.
2	1.5	0	0	.75		21
4	3.5	0	.75	5	.4	12
3	-1.5	4	.4	2	.5	21
2	-1.5	3	.5	.6		22
3	-3	-6.5	.6	7	.5	22
2	1	.5	.5	0		22
0						

X Parity = 0   Y Parity = 0

Figure 5

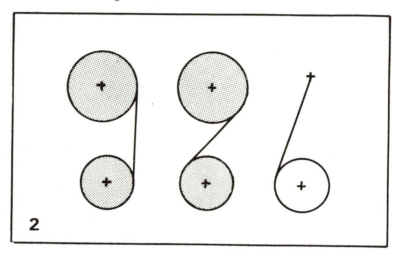

Figure 1   Type 2 Composite

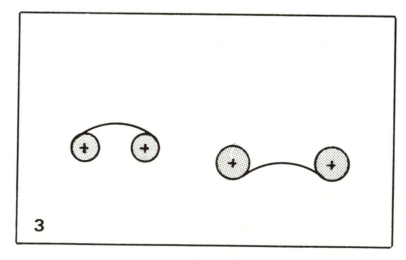

Figure 2   Type 3 Composite

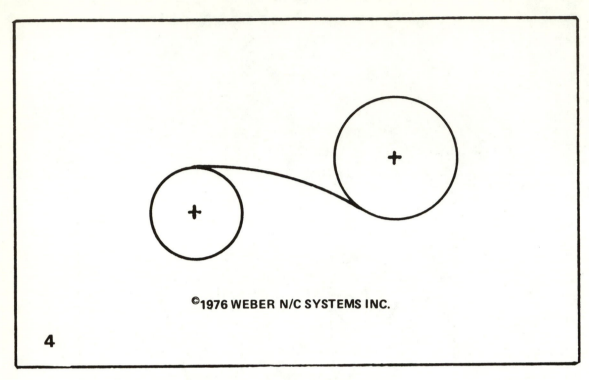

Figure 3   Type 4 Composite

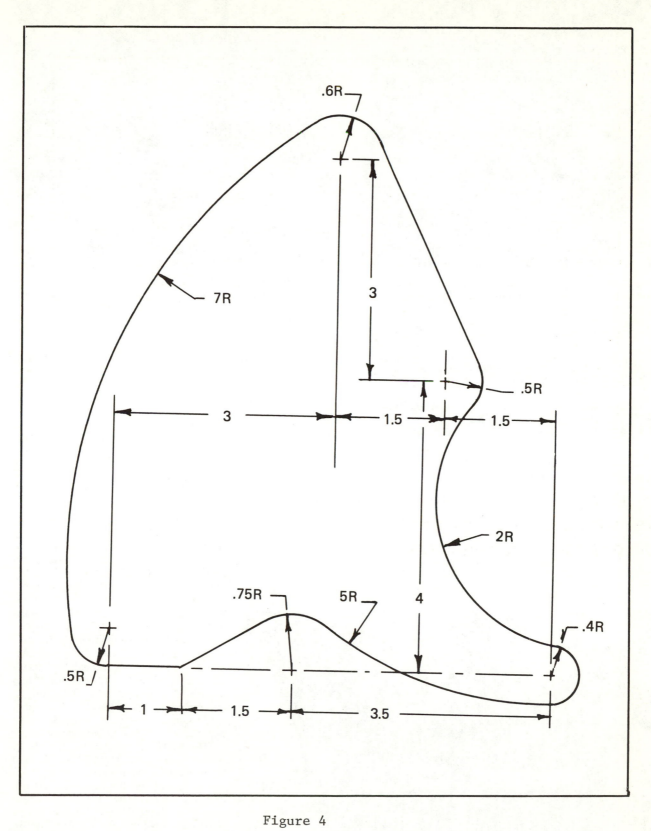

Figure 4

FOUR-AXIS NC turning machines have cut costs up to 40% over the two-axis machines used previously.

TYPICAL PARTS that are turned, grooved, threaded, contoured, and bored on the four-axis NC machines.

Reprinted from: Manufacturing Engineering, November 1979

Cutting Costs by 40% with Four-axis NC Turning

A switch to double-slide turning machines has allowed this oil well equipment maker to combine cuts, and use a permanent tooling system for families of parts

END TURRET on four-axis NC turning machine provides necessary support for shaft-type parts.

IN THE LAST 10 YEARS, the oil well drilling equipment industry has had one of the most dramatic growth records in the U.S. Sharing in this bonanza is Gearhart-Owen Industries, Inc., Fort Worth, Texas, whose sales zoomed from about $2 million in 1961 to over $100 million in '78. The firm manufactures sophisticated downhole analyzation equipment principally used to log activity such as oil, gas, and water levels and temperatures.

An important part of GOI's production operation is the turning of metal parts. Most of these components have multiple diameters, and require grooving, threading, contouring, and boring, as well as basic turning operations. Materials machined include 17-4 stainless steel (from 40 to 50% of the total), high carbon steels (4340 and other alloys), cast iron, and some brass and aluminum alloys. Most of the high carbon steels are heat treated to a hardness of R_C 35-40. Tolerances range from a common ±0.002" (0.05 mm) to an occasional ±0.0005" (0.013 mm), and surface finishes are generally in the 32 to 63 μin. (0.8 to 1.6 μm) range.

The Switch to Four-axis. Until 1970, this firm used two-axis NC machines to turn most of these parts. But then, in order to cut machining costs and improve flexibility, their production management team decided to switch to four-axis turning. As George Smith, GOI's plant superintendent, states, "It just made sense to try to get the most cutters on the workpiece we possibly could. We wanted to combine cuts whenever possible."

To fulfill this commitment, the firm selected Warner & Swasey four-axis double-slide, slant bed NC turning machines. The first machine went into production in July, 1970. Today, the shop has a lineup of 11 machines — nine SC-15's and two SC-25's. Four of the machines have Allen Bradley 7340 CNC controls. At present, tapes have been prepared for almost 4000 different parts, representing 80 to 90% of all components the firm turns.

Benefits Obtained. Overall, the conversion to four-axis turning has resulted in a 30 to 40% savings over costs with the previous two-axis machines. Obvious savings came from the ability to take combined cuts on the four-axis machines. In addition, when only one turret is cutting, the other is independently indexed and positioned for the next cut, thus minimizing idle times.

Cost reductions have also resulted from the versatile permanent tooling system which uses standard off-the-shelf, qualified cutters in all 11 tool positions. And, because of the end turret's ability to cut on both sides of center, additional ID machining capability is available by using double cutter boring bars.

Most machines now stay permanently tooled for entire part families, enabling the firm to reduce tooling and job changeover costs. Inventory levels can now be kept much lower, because it's even economical to run lots as small as five parts.

Both chucking and shaft work can be done on the same machine because each turret operates on an independent slide. For long shafts, the end turret provides the necessary end support to minimize chatter and maintain close tolerances. In most cases, secondary grinding operations are not needed.

George Smith states, "We learned early that we could handle all 10 sizes of our shaft-like bridge plug bodies without tearing down an entire end turret, ID tooling arrangement from a previous setup. By simply substituting a revolving center support for one end working tool, we were able to run off our first lot of bridge plugs with no special tooling or holding equipment."

The 20° slant bed design of these machines allows easy operator access for fast part loading/unloading, gaging, and tool changes. Large diameter CURVIC couplings on these double-slide machines provide repetitive indexing accuracy, and permit performing roughing and finishing cuts in one operation.

Programming of these machines is easy, partly because of the ample tool clearances between adjacent stations, and their independent turret operation. Another benefit realized from the installation of these machines is that they have helped solve the skilled labor shortage problem. With most jobs pretooled by a setup man, it's easy to train machine operators. ∎

Programming NC Traveling Wire Machines
Numerical Control Programming Process

By Raymond J. Gaynor
Corporate Applications Engineer
Manufacturing Data Systems, Inc.

This paper will give an overview of the entire Numerical Control programming process, beginning with the basic question "What is a Numerical Control program?" Discussions on programming methods available, manual and computer assist, with examples of both, will follow. It will continue with an introduction of the critical factors of accuracy, productivity, cost-effectiveness and lead time for determining the programming method. Programming examples will emphasize computer assist methods. The paper will also consider program debugging and verification.

INTRODUCTION

The NC programming process begins with the part programmer studying the workpiece drawing. In effect he is much like a machine operator, studying the part drawing before starting the machine. However, he will not turn handwheels and operate controls as would the machine operator. The programmer visualizes the machine motions necessary to machine the part. Once conceptualizing them, he will document them in a logical order. This information is then translated to a programming manuscript. He will do this whether the program itself is to be developed manually or with computer assist. The program is then checked by plotting out the coordinates from the programming manuscript. This may be done manually or via an NC graphic system. Once the programmer is convinced that all his information is correct, the manuscript data is then converted to a medium (usually perforated tape) that can be input to the machine control unit. The programmer now has full responsibility for operating the machine tool effectively.

Programming an NC traveling wire machine is not difficult. Basically it is documenting all machine motions and functions needed to machine a specific workpiece.

Before discussing NC programming we must first understand what a Numerical Control program is. By strict definition Numerical Control is the operation of machine tools by a series of coded instructions, which are comprised of numbers and other symbols. Webster's definition of a program is "a logical sequence of operations to be performed". Coded commands, gathered together and logically organized so that they will direct a machine tool in a specific task, comprise an NC program. In other words, it describes tool locations relative to the workpiece or the machine tool coordinate system.

Machine tools themselves cannot read coded commands. The actual reading is done by a program reader, usually a tape reader, which is part of the machine control unit (MCU). The MCU converts the coded instructions into output signals which control the electric, hydraulic or other type of servomechanisms used to physically drive and direct a machine tool.

Conventions for the communication to the machine tool of the workpiece and tool locations must be established between the programmer and the MCU. This is usually accomplished through the Cartesian coordinate system, also known as rectangular coordinates (Fig. 1), in specifying locations. The concept fits machine tools perfectly. Machine tool construction is normally based on two or three perpendicular axes of motion. It is the universally accepted standard. The system is defined by a set of three mutually perpendicular planes, with the common point of intersection of the three planes called "absolute zero". The intersection of each pair of planes forms a line in space referred to as the X, Y, or Z-axis. An easy way to remember this convention is the "Right-hand Rule" (Fig. 2). For traveling wire machines we will only use two planes, since there is no up or down tool motion (Z movement). These two planes will form an X and Y coordinate system.

In addition to describing the tool locations with X and Y coordinates, we must also describe other pertinent information to the MCU, dealing with rate of motion and type of motion, i.e., linear or circular interpolation, feed or rapid.

Whereas motion is coded in X and Y locations other functions may also be coded with G and M or other codes, each with a function number. These additional instructions are called Preparatory and Miscellaneous functions (Fig. 3).

Programming Methods

There are basically three types of programming methods to choose from: manual part programming, batch entry computer assist, and computer assist with interactive processing.

Manual programming forces the programmer to calculate the tool locations for each point, a new point being required every time the tool path changes direction. The programmer must also add the preparatory and miscellaneous functions in the proper sequence.

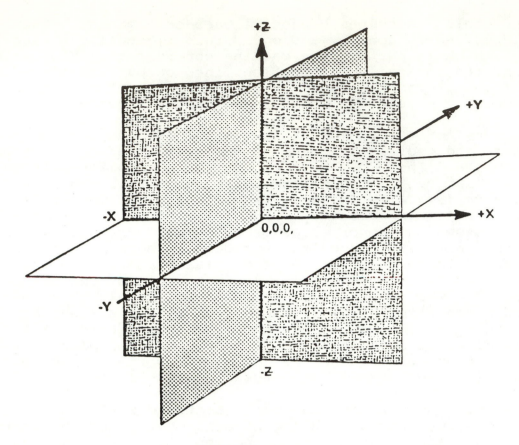

FIGURE 1

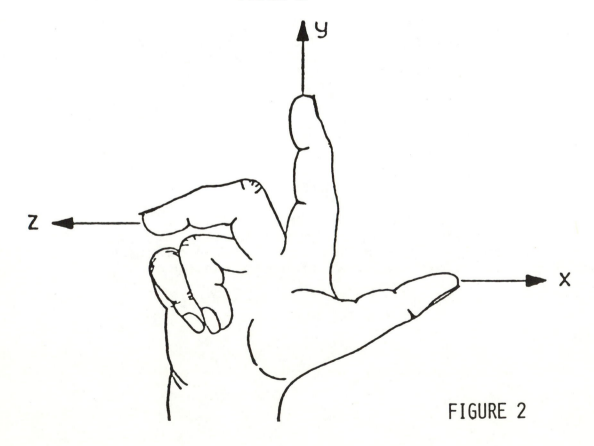

FIGURE 2

CODE	PREPARATORY FUNCTIONS
G00	POINT TO POINT POSITIONING
G01	LINEAR INTERPOLATION (NORMAL DIMENSIONS)
G02	CIRCULAR INTERPOLATION ARC CW (NORMAL DIMENSIONS)
G03	CIRCULAR INTERPOLATION ARC CCW (NORMAL DIMENSIONS)
G04	DWELL
G17	XY PLANE SELECTION
G18	ZX PLANE SELECTION
G19	YZ PLANE SELECTION
G40	CUTTER COMPENSATION CANCEL
G41	CUTTER COMPENSATION - LEFT
G42	CUTTER COMPENSATION - RIGHT
G90	ABSOLUTE POSITIONING
G91	INCREMENTAL POSITIONING
G94	IPM MODE
G95	IPR MODE

MISCELLANEOUS FUNCTIONS

M00	PROGRAM STOP
M01	OPTIONAL (PLANNED) STOP
M02	END OF PROGRAM
M03	SPINDLE CW
M04	SPINDLE CCW
M05	SPINDLE OFF
M06	TOOL CHANGE
M07	COOLANT NO. 2 ON
M08	COOLANT NO. 1 ON
M09	COOLANT OFF
M30	END OF TAPE

FIGURE 3

Trigonometry and geometry are used to calculate the tool locations. In the illustration of linear motion, (Fig. 4), you will note that tool location 1, TL#1, is calculated by taking the X-axis dimension and adding the tool radius, giving an overall dimension X3.00. The Y-axis location is calculated in the same manner. The second tool location, TL#2, is slightly more difficult to calculate. First we must calculate angle α in order to determine the σ angle. Once the σ angle is known we are able to calculate distance yl, yl=tool radius x tan σ. Now yl is added to Y-axis dimension of the two intersecting elements. The third location, TL#3, is calculated similarly to TL#2. Now yl is added to the X-axis dimension of the two intersecting elements. Calculating circular motion coordinates becomes somewhat more difficult, as you will note, (Fig. 5). These are the mathematical formulae necessary to determine the start and finish points for radii and filets. This information is then transferred to the programming manuscript. Finally the data from the programming manuscript describing tool locations, rate of motion and other pertinent machine control data is punched into a paper or mylar tape which is read by the MCU.

Computer assist programming systems allow the programmer to describe part shapes, machine tool motions, and machine functions in English-like commands. Using computer assist programming methods, parts may be defined as a series of points, lines, circles, other curves and planes, and the tool path may be directed relative to the defined shape. The computer system compensates automatically for the tool size and shape. It handles all mathematical computations and tape formatting problems, relieving the part programmer of these tedious and error prone activities. Changes can be made quickly, easily and accurately to accommodate unforeseen conditions, to change machine functions, or to add engineering change information. The same program may be reprocessed to produce a tape for a different machine tool to satisfy changing machine loading requirements. The English-like commands for part geometry, tool motions and machine functions are translated by the software into proper coding and tape format for the machine being programmed.

The following sequence describes a typical programming session, developing a program for computer assist. By studying the part drawing, (Fig. 6), the programmer visualizes the machine motions necessary to produce the part. He selects the far left center point of the part to be his start point, from that point he will direct the tool to move in a clockwise direction until he returns to his start position. He will now start to write his source program, (Fig. 7). The first step is writing the initialization section of the program. This section calls up the proper link (postprocessor) for his specific machine. It will also identify the program and the machine tape, establish setup position, and may describe specific types of output desired.

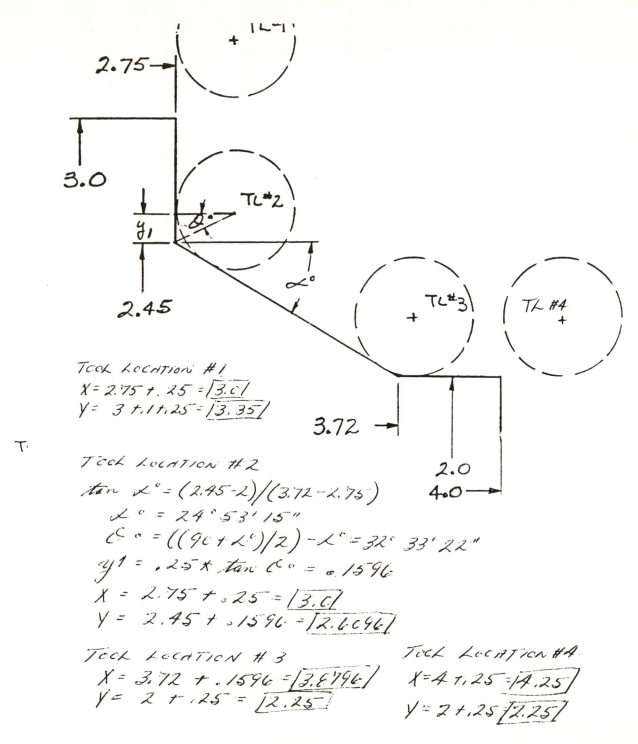

2.75 →

3.0

y_1

2.45

TL#1 +

TL#2

d°

α°

TL#3 +

TL #4 +

3.72 →

2.0

4.0 →

TOOL LOCATION #1
$X = 2.75 + .25 = \boxed{3.0}$
$Y = 3 + .1 + .25 = \boxed{3.35}$

T.

TOOL LOCATION #2
$\tan \alpha^\circ = (2.45 - 2)/(3.72 - 2.75)$
$\alpha^\circ = 24^\circ 53' 15''$
$C^\circ = ((90 + \alpha^\circ)/2) - \alpha^\circ = 32^\circ 33' 22''$
$y^1 = .25 \times \tan C^\circ = .1596$
$X = 2.75 + .25 = \boxed{3.0}$
$Y = 2.45 + .1596 = \boxed{2.6096}$

TOOL LOCATION #3
$X = 3.72 + .1596 = \boxed{3.8796}$
$Y = 2 + .25 = \boxed{2.25}$

TOOL LOCATION #4
$X = 4 + .25 = \boxed{4.25}$
$Y = 2 + .25 \boxed{2.25}$

Fig. 4

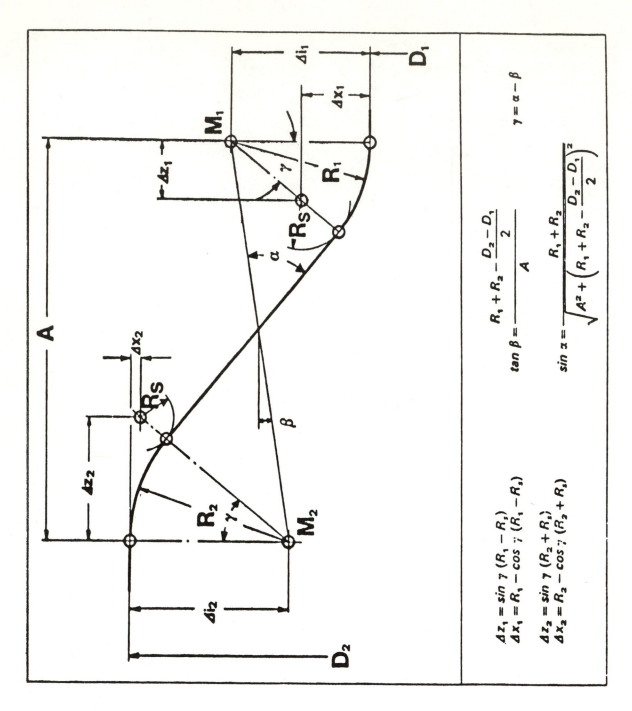

$$\Delta z_1 = \sin \gamma \, (R_1 - R_s)$$
$$\Delta x_1 = R_1 - \cos \gamma \, (R_1 - R_s)$$

$$\Delta z_2 = \sin \gamma \, (R_2 + R_s)$$
$$\Delta x_2 = R_2 - \cos \gamma \, (R_2 + R_s)$$

$$\tan \beta = \dfrac{R_1 + R_2 - \dfrac{D_2 - D_1}{2}}{A}$$

$$\sin \alpha = \dfrac{R_1 + R_2}{\sqrt{A^2 + \left(R_1 + R_2 - \dfrac{D_2 - D_1}{2}\right)^2}}$$

$$\gamma = \alpha - \beta$$

Fig. 5

The next step is to determine what part description parameters will be required. The first element of part geometry the programmer will define is the center of the part, PT1, and since most of the dimensions are taken from this point he will make this his base position. Base is the zero point of a secondary coordinate system. Using the base as a reference point he is now able to describe the geometry with incremental dimensions from "OptiMill" to describe the part and tool motion, a minimal amount of defined geometry is needed. For this part (Fig. 5), you will note only four elements of geometry were necessary, one point and three circles. The programmer then proceeds to define the part boundary. The first element is -.8706 in X from base (-.8706XB) and begins at zero in Y from base (YB). The next element is a .065 radius, then .3562 in the Y direction from base. The .3731 radius is not mentioned at this time, going directly to the .5284 radius or "CIR2". The next element is a line that is tangent to CIR2 and CIR3, on the Y-large side (top of the part), then "CIR3". Another line tangent to CIR2 and CIR3 is next, but on the Y-small side (bottom of part). The programmer then bypasses the .3731 radius and describes a line (LN5) that will be tangent to "CIR4" (.1860R) and is on a 35° angle. Next is "CIR4" then a line that is parallel to LN5 on the X-small side of "CIR4". Again the programmer omits the .3731 radius and moves to -.8706 in X from base, then up to 0y (YB), completing the part boundary definition.

The next step is to describe the tooling section, bringing out information pertaining to the tool, in this case .008dia wire (TD.008). The programmer then describes the motion statements to cut the part. He uses the word CUT, meaning motion at feedrate; PB1/CL which tells the system the tool is to move at feedrate around part boundary 1 with the tool on the left side of the boundary; and by stating .3731R, the system will automatically insert a .3731 blending radius wherever two elements create a nontangent intersection.

The final step to the program is the termination section. This is accomplished with the word "END". It will give the proper codes to return to the start position and terminate all machine tool functions.

At this point the programmer would transfer the information from the handwritten manuscript to paper tape. Using batch processing computer assist the tape would be submitted and the results of the processing would be ready in less than a day, if a computer dedicated to NC is available. Using an inter-active system, the program can be debugged, the tool path plotted, (Fig. 8), and machine control tape produced in a single terminal session of less than one hour. Since the tool path can be plotted prior to producing the machine control tape and carefully scrutinized for programming errors, virtually no prove-out time is required on the machine tool.

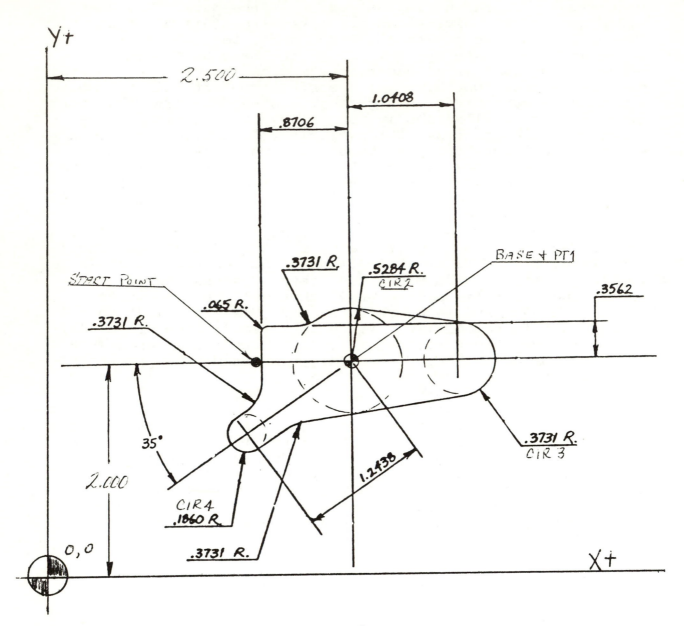

FIG. 6

```
MACHIN, _____

IDENT, M.D.S.I. DEMO - S.M.E.
          EDM CLINIC

SETUP, .75LX, 2LY

BASE, PT1(2.5XA, 2YA)

DCIR2, PT1, .5284R

DCIR3, 1.0408XB, YB, .3731R

DCIR4, PT(PT1, 1.2438R, 145CCW), .186R

DPB1, S(LN(YB)), -.8706XB ; .065R
   ; .3562YB ; CIR2, CW, S(XS)
   ; LN(CIR2, CIR3, YL) ; CIR3, CW
   ; LN(CIR2, CIR3, YS) ; LN5(CIR4, XL, 35CCW)
   ; CIR4, CW ; LN5/.372XS
   ; -.8706XB, F(LN(YB)), NOMORE

ATCHG, TD.008

CUT, PB1/CL, .3731R

END
```

Fig. 7

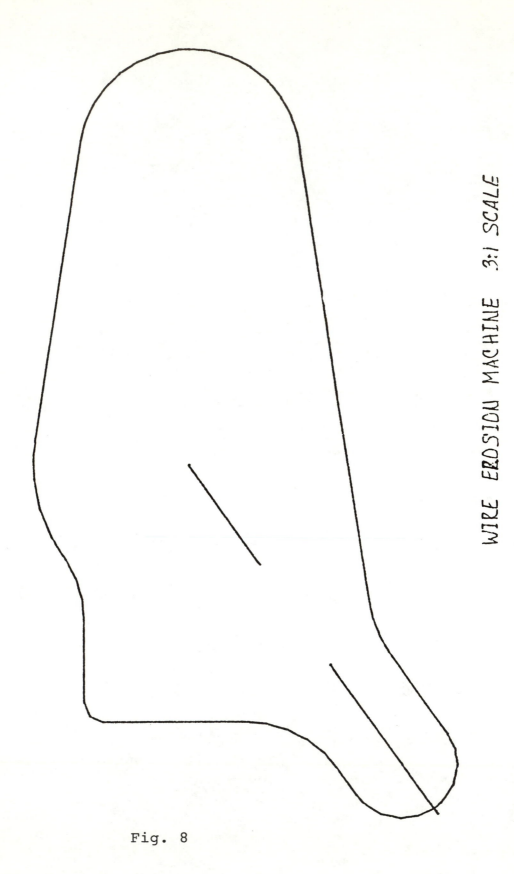

WIRE EROSION MACHINE 3:1 SCALE

Fig. 8

The following examples show typical parts suited for the traveling wire machines. In Figure 9 you will note that the part actually consists of a very simple shape. The difficulty occurs when this shape is rotated around a common point, at different angles, nine times. By using the computer assist method of programming, (see Fig. 10), the programmer is able to describe the motion once and then rotate that motion the remaining eight times, leaving the computer to calculate the rotated locations.

In Figure 11 we have again a very simple shape which is duplicated four times at different locations. Figure 12 shows how we are able to describe the shape and motion once and then translate it to the proper locations. Also note that the print calls for all intersections to have .008 radius, which is accomplished by adding a blend radius value to the "CUT,PB1" statement.

Some critical factors to consider when choosing the programming method are: accuracy, productivity, cost-effectiveness and lead time. The system offering the highest accuracy, fastest turnaround, and fewest man-hours for the intended application is usually the most productive.

Cost-effectiveness of the machine control tape is dependent upon the programmer involved. A person with good "machine knowledge", an understanding of layout procedures, set-ups, and basic math, usually will make a good programmer.

Lead time for machine control tapes are usually directly related to the programming method or system. In today's highly competitive market no NC machine can afford to sit idle waiting for a tape. If this should happen at your facility a reevaluation of programming methods is indicated.

A major factor in the final selection of the method or system should be the results of comparative demonstrations of capability and time required using one or more actual parts to be manufactured. If several computer assist systems are to be considered, and they should be, insist on a start-to-finish demonstration in your shop, on your parts. Also request the names of shops in your area that do similar work or have similar machine tools, whom you may call for a recommendation.

The decision to use either computer assist or manual programming methods for tape production should be based upon the configuration or other factors relating to the parts or tools to be produced and the types and variance of the machines to be programmed, and not necessarily upon the number of parts or machines to be programmed.

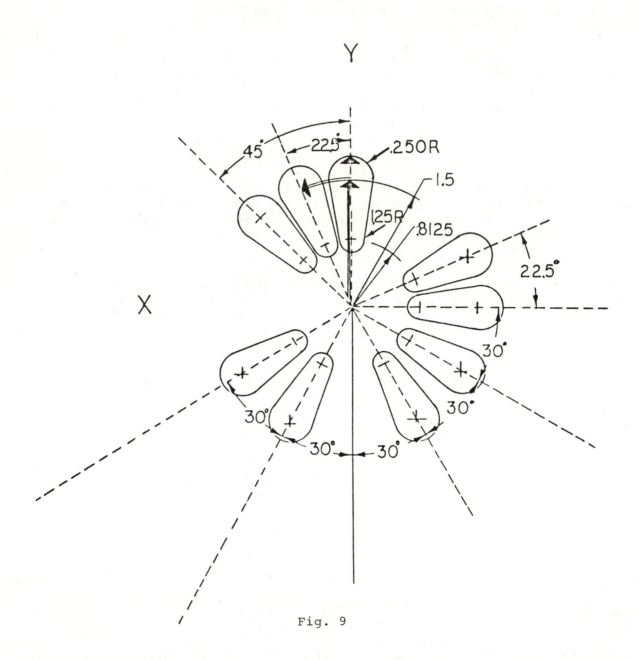

Fig. 9

```
MACHIN,
IDENT,SHAPE #5
SETUP,LX,LY
DPT1,XB,1.5YB
DCIR1,PT1,.25R
DCIR2,PT1,.8125YB,.125R
DLN1,CIR1,CIR2,XS
DLN2,CIR1,CIR2,XL
MTCHG,.005TD
DVR1,0
DVR3,22.5
<1>MOVE,PT1,STOP
CONT,CIR1,90CCW,F(TANLN1)
 ;CIR2,S(TANLN1),F(TANLN2)
  ;CIR1,CCW,S(TANLN2),F89,STOP        $GLUE STOP
  ;CIR1,CCW89,F90
MOVE,PT1,STOP
<2>DVR1,#1+#3
DO1/2,ROTXY-#1,2TMS
DVR3,30
DVR1,120
DO1/2,ROTXY-#1,2TMS
DVR1,210
DO1/2,3TMS,ROTXY-#1
DO1/2,ROTXY67.5
END
```

Fig. 10

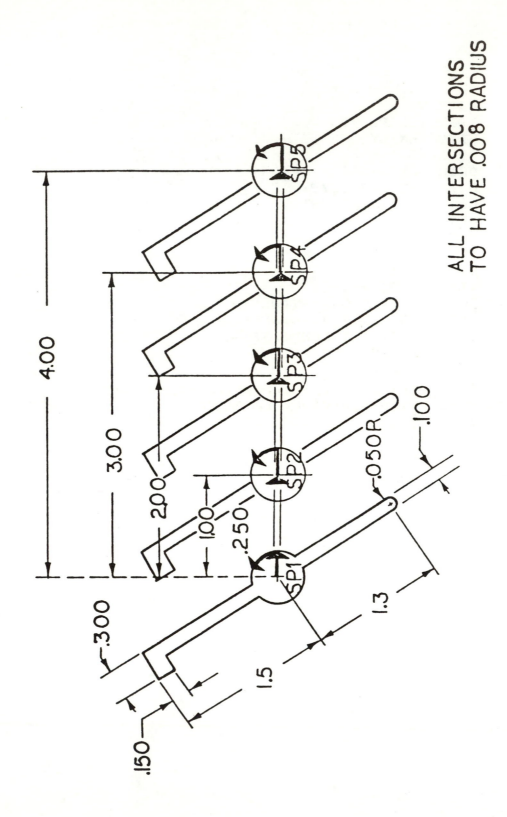

ALL INTERSECTIONS
TO HAVE .008 RADIUS

Fig. 11

70

```
MACHIN,
IDENT,SHAPE # 8
SETUP,LX,LY
MTCHG,.008TD
DCIR1,LN1(XA,YA,60CW),CIR2(XA,YA,1.3R),YS,.05R
DLN2,LN1,CIR2/.2,YL,PERLN1
DPB1,ON,RPT,XA,YA;SPT1,XA,YA;CIR3(XA,YA,.25R),S(LN(YA),XL),CCW
  ;LN1/.05XL,S(YL);LN2;LN1/.25XS;LN2/.15YS;LN1/.05XS
  ;CIR3,CCW,S(YL);LN1/.05XS,S(YS);CIR1,CCW;LN1/.05XL
  ;CIR3,CCW,S(YS),F(LN(-.01YA),XL),NOMORE
DVR1,0
<1>MOVE,PB1/SPT1,STOP
CUT,PB1,.009R,STOP        $ GLUE STOP
CONT,CIR3,CCW,S(LOC),F0
CUT,PB1/SPT1,STOP
<2>DVR1,#1+1
DO1/2,#1X,4TMS
END
◆
```

Fig. 12

NC Punching Machines. . .Programming Profitability Through Tapes

By Leonard A. Weibel
NCL Product Manager
Manufacturing Data Engineering Services Division
Structural Dynamics Research Corporation

Numerical Control (i.e., NC) techniques have provided users with greatly increased production flexibility. NC programming allows the machine operator to stand away from the machine and convey metalforming information to it by way of tape. The challenge is to produce NC machine tapes which are both efficient and effective as regards the programmer and the machine tool. NC programming includes areas of economic opportunity for programmer time; graphics; programming aids; family parts; sheet layout; machine cycle optimization; and first part proveout.

What does the future hold in store for the typical job-oriented, metalforming shop. For openers, lower priced programming systems; expanded graphics and programming aids; direct Numeric Control; and manufacturing management systems.

INTRODUCTION

Civilized man has lived on earth for more years than history has recorded. However, man's development as a productive creature has developed rapidly. Consider, for example, that man started working with brass and other metals barely 6,000 years ago. Let's assume that our civilization's last 10,000 years on earth represent only one year and offer some perspective on our growth.

- 1 year ago civilized man appeared (8000 B.C.)
- 7 months ago man started working in brass (4000 B.C.)
- 3 months ago Euclid organized geometry (300 B.C.)
- 21 days ago the Renaissance began (1400 A.D.)
- 9 days ago the Industrial Revolution started (1740)
- 8 days ago John Wilkinson built the first practical machine tool (1774)
- 81 hours ago Nikola Tesla developed practical AC Power Systems (1888)
- 65 hours ago William Sellers developed the first 36 tool automatic tool punch press (1906)
- 31 hours ago Presper Eckert and John Mauchly developed ENIAC computer (1944)
- 25 hours ago MIT developed first NC machine tool and APT language (1952)
- 9 hours ago Neil Armstrong first walked on the moon (1969)
- 4 hours ago CNC machine tool control units were introduced (1975)
- 52 minutes ago minicomputers have become the wave of the future (1979)

In a sense, we are approaching the 21st Century with blinding speed. The future belongs to those fortunate few who have the capacity to visit the past and present and take away from it their future success strategy. This paper deals with proftability available to NC Sheetmetal Shops by way of programming. All shop activity is touched by NC.

BACKGROUND

The 12th American Machinist Inventory of Metalworking Equipment 1976-78, contains some interesting facts about NC punching and shearing. Consider the following:

MACHINE AGE	GROUP TOTAL	PERCENT
0-4	8,265	13.9
5-9	14,830	24.9
10-19	21,683	36.4
20 & Over	14,825	24.8
TOTAL	59,603	100.0

In this group, only 1,522 machines or 2.6% are NC equipped. Also, over 60% of the reported machines are 10 years or older. It is clear that this area will receive much attention in the next 5 years as the need to improve sheetmetal productivity grows.

Machine Tool Productivity as reported by American Machinist has increased dramatically over the past 10 years in relation to the number of machine tools sold (see Figure 1). This is largely influenced by NC machine tools because during this same period the number of NC machine tools sold has grown substantially over the number of non-NC machine tools sold (see Figure 2).

A recent SDRC report indicated that over 50% of all NC punch press equipment is programmed manually, that is to say, without the aid of any computer. The remainder are programmed by over fifteen (15) different vendors offering timesharing, large in-house mainframe or minicomputer systems. A list of these vendors may be found in Figure 3. Note that many NC programming vendors only support one or two distribution channels (i.e., Timeshare, In-house or Minicomputer) and that others only support their own punch press equipment. One purpose of this paper will be to review the merits of manual vs. NC, computer-assist sheetmetal programming and then to offer criteria on which to evaluate computer-assist alternatives.

DEFINITIONS & CONCEPTS

Consider the following when relating numeric control to machine tools:

NC Machine Tools

- Machine tool CONTROL is shifted from the operator to the Machine Control Unit (i.e., MCU). Metalforming is not directly effected by NC.
- NC machine tools are more efficient than conventional machine tools because of higher production rates; improved repeatability and accuracy.
- NC machine tools are more effective because various operations are now combined on one machine setup (e.g., punch press for punching, shearing, etc.)

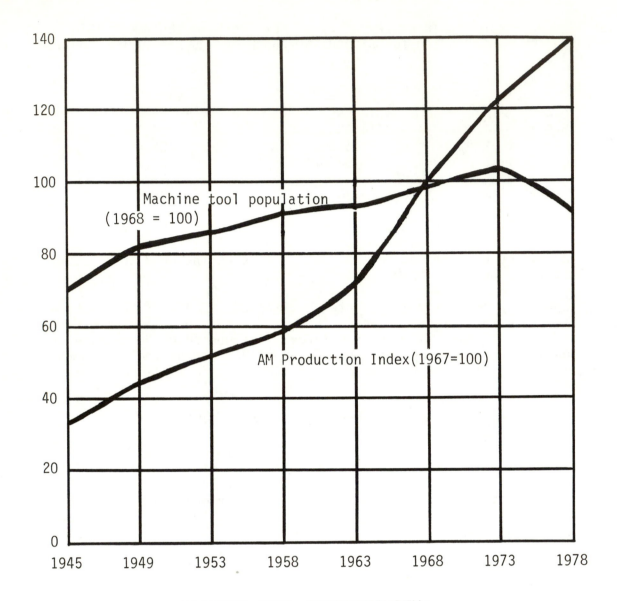

MACHINE-TOOL PRODUCTIVITY:

MORE PRODUCTION FROM FEWER MACHINES

FIGURE 1

Source: 12th American
          Machinist Inventory

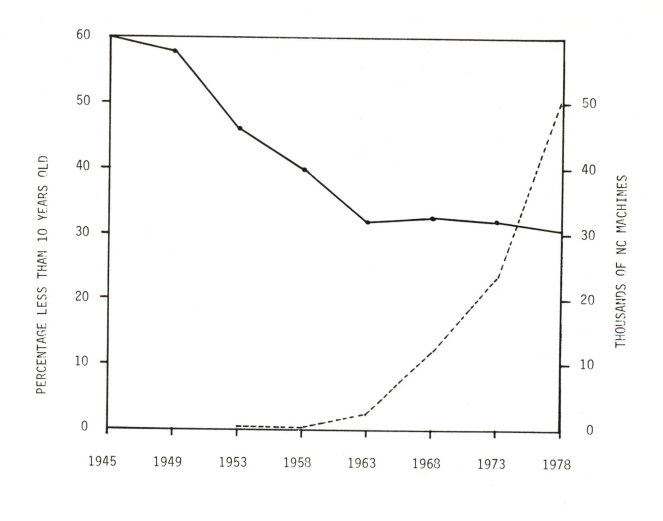

FIGURE 2

Source: 12th American
Machinist Survey

VENDOR	LANGUAGE	AVAILABILITY		
		TIMESHARE	IN-HOUSE	MINICOMPUTER
Amada ① ②	Optiplot ™			X
Data General ②	Dataprep ™			X
Encode	Genesis ™			X
General Electric	NCPPL ™	X		
LeBlond, Inc.	LTTP ™	X	X	X
MDSI	Compact II ™	X	X	X
Marko Systems	Quick Disc ™			X
SDRC	HI-PRO ™	X	X	X
Strippit ① ②	Fabpoint ™			X
Techware ②	Cam Tech ™			X
Threshold Technology	VNC-200 ™			X
UCC	APT	X	X	X
Weber Systems	PROMPT ™			X
Westinghouse Electric	APT	X	X	
Wiedemann ① ②	Wiedepoint R			X

① Programming System designed to support only that specific vendor's equipment.
② Punch Press Programming only.

Figure 3

NC Part Programming

- A part program is a specifically sequenced combination of machine tool motion and function commands which are used to direct the machine tool through its MCU in the punching of a part.
- The part programmer's intellect and experience now replace that of the trained machine operator.
- An experienced programmer can extend his shop experience simultaneously over many NC machine tools as contrasted with only over one or two conventional machine tools.
- The key to sustained high utilization of NC machine tools is a backlog of efficiently prepared and operating machine tapes.
- Regardless of the methods used to prepare an NC part tape (i.e., manual or computer-assist), the end product must contain specific information.

Types NC Programming

- Point-to-point entails movement to discrete locations on the part with the cutting tool NOT in contact with the workpiece except for the actual operation (e.g., punching, etc.)
- Continuous Path (i.e., contouring) involves movement to discrete locations on the part with the cutting tool in CONTINUOUS contact with the workpiece for the actual machining operation (e.g., turning, milling, burning, etc.)

MANUAL NC PART PROGRAMMING

A manually prepared part program represents a considerable amount of clerical labor wherein many opportunities for error exist. We have chosen a straightforward, demonstration shaker part to contrast manual part programming with computer-assist techniques. Figure 4 represents the part in its simplest configuration and that which we will use to illustrate some fundamentals of manual, punch press programming. Our objective is to write a punch tape manuscript adequate for use on the STRIPPIT® FABRI-CENTER® 1000 equipped with Houdaille 1000A Machine Control Unit. This punch press is representative of incrementally programmed machine tools. That is to say, every table movement command is given in relation to the previous command.

Machine Orientation

In order to prepare a program manuscript for a part, a programmer must first orient himself with the geometry of the machine tool with which he will be working. Generally, the X-axis "+" moves from left to right when facing a punch press and the Y-axis "+" moves from front to rear. However, differences can occur with regard to where zero is in this two-axis coordinate system and alter our concept of "+" and "-" axes direction. For example, zero may be in the lower left-hand corner or quadrant one; it may be in the upper right-hand corner or quadrant three or for that matter could be located according to various machine tool builder conventions in either quadrants two or four. Of course, the confusion becomes greater when a part programmer is required to support work across several different makes, models and types of machine tool because of the axes orientation difficulty suggested above. Not only are axes orientations different, but also workholder dimensions, repositioning movements and side-plate positioning for part zero designation.

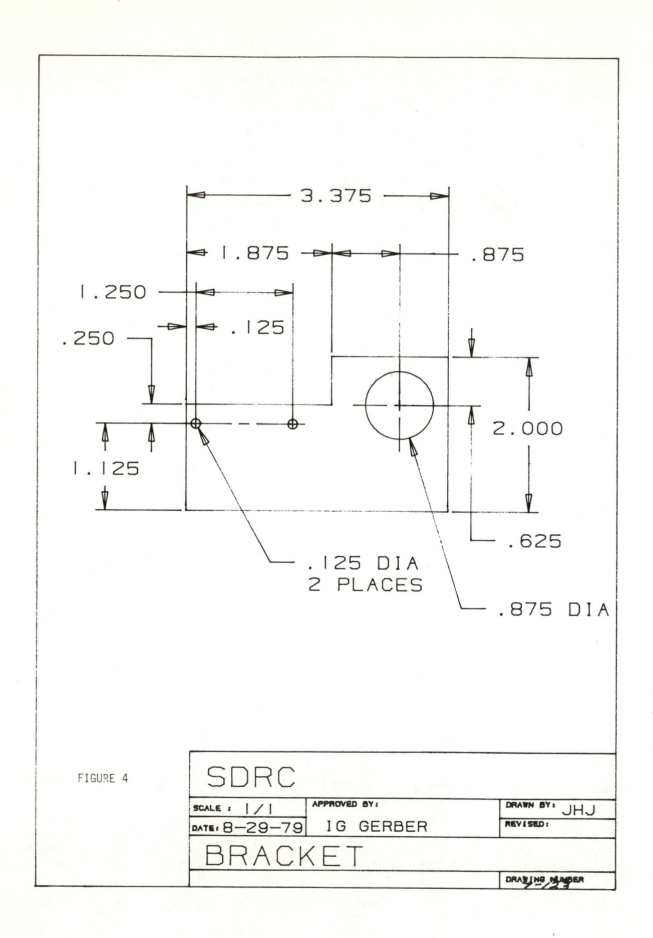

FIGURE 4

SDRC		
SCALE : 1/1	APPROVED BY:	DRAWN BY: JHJ
DATE: 8-29-79	IG GERBER	REVISED:
BRACKET		
		DRAWING NUMBER 7-129

3.375

1.875 · .875

1.250

.250 · .125

.125 DIA
2 PLACES

1.125

2.000

.625

.875 DIA

The definition of part programming, given earlier, discussed the integration of machine codes and geometry in the MCU in order to successfully direct the machine tool throughout each moment of its operation at each coordinate work location. There are "G" (General) Functions, "M" (Miscellaneous) Functions, "S" (Stroke) Functions, "F" (Feed) Functions, "T" (Tool) Functions, etc. Each of these are generally of the form NN (e.g., G90) where the function is a single letter mnemonic plus a numeric value ranging from 0 to 99 as shown previously.

There are many different coding conventions used between manufacturers of different machine tools and MCU's. These combinations of codes are not consistently applied between combinations of machine tool and MCU manufactured by the same firm, let alone among different firms. Over the years, many machine tool builders have used different makes and models of MCU in conjunction with many makes and models of machine tools. Additionally, special machine tool options can cause variations in MCU codes between even similar makes and models of machine tool and MCU. This has led to a great deal of variation and causes the part programmer a great amount of difficulty when he must support many differing coding combinations, again under the pressure of an on-going shop operation.

Program Coding

In order to prepare a program manuscript for a part, a programmer must first translate the dimensions found on the part layout print to incremental machine axes movements by resorting to addition, subtraction, multiplication, division, right-angle trigonometry, etc. Unfortunately, part prints are not drawn incrementally and during this operation, the possibility of error(s) becomes important. Consider, for example, any axis error will be propagated throughout the remainder of all hits. Of course, additional errors only compound the problem.

The manual, NC part programmer writes a line or block of program data on the coding sheet, which represents a single, machine motion (motions). As shown in Figure 5, every motion or non-motion command must be accounted for carefully. If not prepared carefully, the results can vary from no apparent effect through that of creating a defective piece-part or worse, a machine crash. This particular example requires approximately 19 blocks of programmed data in order to complete the part as shown. The 19 blocks of data requires one page of coding.

Pattern Layout

In metalforming work, it is many times desirable to propagate a single part across a sheet. In so doing, you greatly increase the amount of work required by the manual part programmer (see Figure 6). The programmer must now effectively write as many additional part programs as there will be parts on the sheet. This is necessary because each part will have a different set of incremental moves describing axes motions to the punch press. You might say that a CNC control unit would automatically propagate the part for the programmer across the sheet, thereby, releasing him of this responsibility. This is so and works well for only the very simplest of parts. However, if, for example, a part requires six tool changes, this feature will punch the part with six tool changes; punch the next part with six tool changes, etc. This is obviously not an optimal fashion in

BLOCK N	PREP CODE	±	X	±	Y	MISC. CODE	TOOL CODE	REMARKS
								INCREMENTAL MODES
N001	G69					MO6		INITIALIZE
N002		01		01		M75		LOAD POSITION
N003	G68	46	625	37	375		T2	1HIT T#2
N004		00	281	-01	369		T8	2HIT T#8
N005				-01	25			ONVERT LINE
N006		-00	593	01	432		T5	1HIT T#5
N007		01	937	00	312		T11	2HIT T#11
N008				-00	4			LTVERT PART LINE
N009		-02	375	01	4		T20	2 HIT T#20
N010				-00	5			VERT PART LINE
N011		00	625	-01	9			2HIT T#20
N012				00	5			VERT PART LINE
N013		01		-01	125		T19	2HIT T#19
N014		-00	375					HORIZ PART LINE
N015		00	375	03	65			3HIT T#19
N016		-00	5					HORIZ PART LINE
N017		-00	5					HORIZ PART LINE
N018	G67	-46	5	-33	125	M75,M71		PUNCH OFF
N019								HOME & REWIND

MANUAL TAPE PROGRAM
DEMO PART BRACKET
FIGURE 5

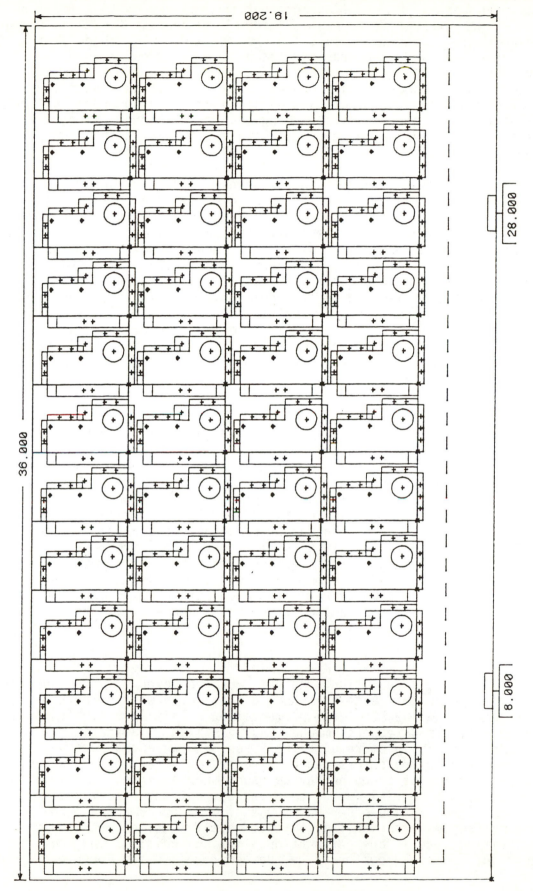

DEMO BRACKET PART SHEET LAYOUT

FIGURE 6

which to punch parts. Figure 12 points out the fallacy of this logic on multiples. The alternative suggestion might then be made for those using CNC machine tools. That is to say, the programmer should schedule each hit for the single part and then propagate each by tool across a sheet by use of the various in-build grid commands. This does work better than the prior alternative. However, part program time still is substantially extended because of the additional consideration required in order to locate distance between centers and establishing the grid pattern. Also, with this situation, any change whatever will cause a change in the overall program tape.

The programmer now has two additional concerns. The first being that of sheet repositioning if the piece-part is larger than the effective work area for the machine on which the part is being manufactured. The part programmer must reschedule all these hits in concert with the piece-part shift. The programmer must, secondly, be concerned about the possibility of hits existing in a clamping, no-hit area. Generally, the machine will inform him of this by either hitting the clamp area, thereby, deforming the sheet or shutting down with an error condition indicating that machine limits have been exceeded. At this point, the operator will inspect the part and probably call the part programmer to the production floor for a conference. The part programmer can either fix the part on the floor or take the part off the machine and have another part run until the first part program can be fixed.

Optimization

The shop-smart programmer will typically schedule the hits in his mind, based on the experience he has developed. This will usually result in a reasonably well optimized tape. However, even here there is room for significant improvement. Just taking all hits on a tool sequence basis doesn't assure optimal scheduling and sequencing of hits across a sheet. Figure 13 represents optimization performed merely by sorting tool hits, whereas, Figure 14 represents the computer further resequencing hits and thereby, provide an additional 18.5% cycle time. Other problems arise when the part programmer discovers that various hits should have been taken prior to hits already coded. He must now physically resequence the part program and his machine tape to reflect the updated resequencing. Here, for example, is where simple data editing errors can easily enter the scene.

Tape Perforation

Following manuscript preparation, the data must be keyed on a tape perforation device in order to create the punched tape which will carry the program instructions to the MCU for implementation. This is a highly routine, clerical operation and requires manual typing skills. Validation of this step is carried out by proofreading the keyed data on a character-by-character basis. Keying errors can be remedied with varying degrees of effort depending on the encoding device used. For example, if a teletype ASR 33 is used, the initial device cost is low, but the effort required to use it is high. On the other hand, use of a micro-computer based NC tape terminal represents high initial cost, but low use effort in areas of MCU tape editing, etc. Tape perforation, listing, editing, merging, etc. are operations generally performed on punched tape and can consume a great deal of the part programmer's time if a slow speed device is used. For example, printers operate from 10 characters per second up to 150 characters per second. Tape punching can be carried out from 10 characters per second up to 75 characters per second. Tape reading can be carried out from 10 characters per second up to 300 characters per second.

Tape Proveout

Following tape preparation, the tape must be proven by mounting it on the MCU and operating the machine in order to produce an inspection quality part. On newer machines, a dry cycle mode allows quick cycling of a tape and a high speed cycling of all machine motions contained on the tape. A quick eye is required to spot any of but the smallest of errors. Only the most blatant errors appear at this time. The moment of truth arrives when the first part is punched and inspected. In this mode, a machine operator and programmer are generally involved with the total cost which can easily exceed $100 per hour.

More times than not, the first pass, in fact several additional passes also, are lost in this verification process because it is simply impossible to accurately desk check work and thereby, create an inspection quality first part. More than likely, the programmer will come back to the machine tool several times and consume substantial time and machine resources.

Sources of Manual Programming Error

Let's take a moment to summarize all the various sources that come together to plague a manual part programmer and thereby, greatly extend part programming costs and timeliness.

1. Misreading part prints due to inadequate training, carelessness, or inconsistent drawing standards.
2. Machine axes disorientation due to the plethora of machine tool and MCU code combinations on the market today.
3. Mathematical errors induced by using the hand calculator incorrectly, misreading algebraic values, transposing numbers or misplacing decimal points.
4. Machine coding errors, e.g., an M03 may be used in place of an M04, whereas, on another machine that particular combination code might be adequate.
5. Clerical errors arise due to carelessness, fatigue or general disinterest in this particular aspect of work.
6. Typing errors are the next logical problem that arise and typically are due to a lack of concentration and manual dexterity.
7. Machine tape (data) editing errors arise when the error is compounded due to an improperly applied correction.

NC COMPUTER-ASSISTED
PART PROGRAMMING

An NC, computer-assisted, part programming system, such as the SDRC HI-PRO, greatly reduces the number of programming lines required to be written by a programmer; points out potential and real errors; and finally, minimizes machine cycle time. It accomplishes this by virtue of previously programmed logic designed to facilitate work reduction, error detection and machine tool cycle minimization.

Language

The complete, NC, computer-assisted part programming system employs English-like sheetmetal terminology to describe part geometry and punching operations. The fundamental elements of any sheetmetal language are the LINE, GRID, BHC, ARC and PT commands as shown in Figure 7. These are usually followed by more complex, automatic profiling or nibbling commands for the generation of multiple hits with minimal data input as shown in Figure 8. The objective of these various commands is to minimize the amount of programmer exposure to potential errors. It is only logical to expect that compounding of basic commands can provide many powerful combinations of sheetmetal geometry. Consider the following combinations:

PRIMARY COMMAND	SECONDARY COMMAND
ARC	ARC
BHC	BHC
GRID	GRID
LINE	LINE
HPAT	HPAT
DPAT	DPAT
	PARC
	PLINE
	CIR
	HOLE
	TRIM

The ability to offset hits for bendline compensation and to otherwise move hits across a sheet (i.e., TRANSLATE, ROTATE, TRANSFORM and MIRROR) now provide a complete set of programming tools to develop sheetmetal part programs.

Many times sheetmetal part geometry is found repeated. During this situation, it would be convenient for the programmer to recall a previously developed hole pattern instead of rewriting the entire pattern each time. Consequently, patterns are generally recordable for later recall and manipulation as required.

Computer-Assist Part Program

We have developed a part program to demonstrate some of the aspects of the SDRC HI-PRO computer-assist, part programming language. The program shown below in Figure 9 will, after having been processed by the SDRC HI-PRO System, produce a machine tape similar in content to that found in Figure 5. This tape, or its equivalent, is required to make the part shown in

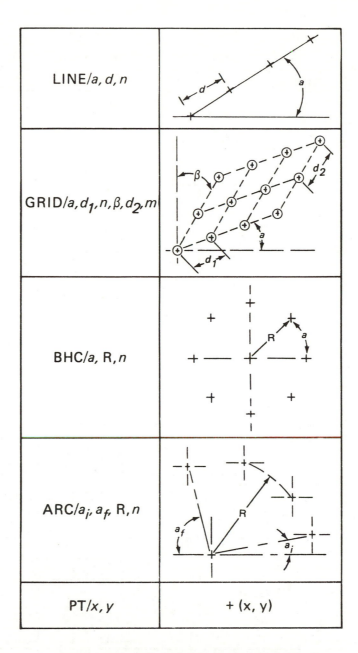

LINE/a, d, n	
GRID/a, d_1, n, β, d_2, m	
BHC/a, R, n	
ARC/a_i, a_f, R, n	
PT/x, y	+ (x, y)

FUNDAMENTAL LANGUAGE ELEMENTS

FIGURE 7

COMMAND	ACTION
PARC/a_i, a_f, R $\begin{Bmatrix} \text{INTL} \\ \text{EXTL} \end{Bmatrix}$, BURR, b	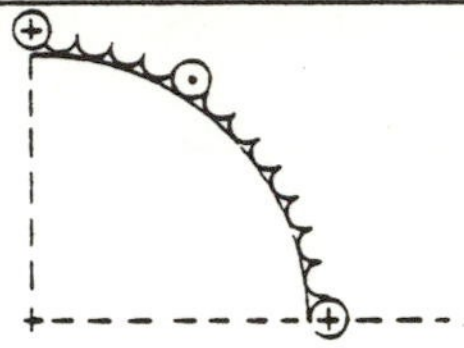
CIR/R, BURR, b, $\begin{Bmatrix} \text{INTL} \\ \text{EXTL} \end{Bmatrix}$	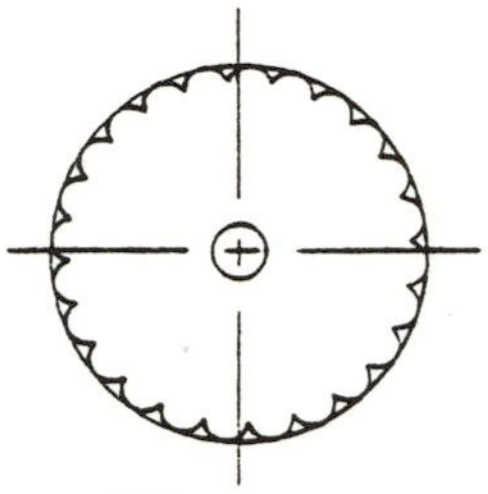
PLINE/a, l, $\begin{Bmatrix} \text{BURR}, b \\ \text{overlap} \end{Bmatrix}$	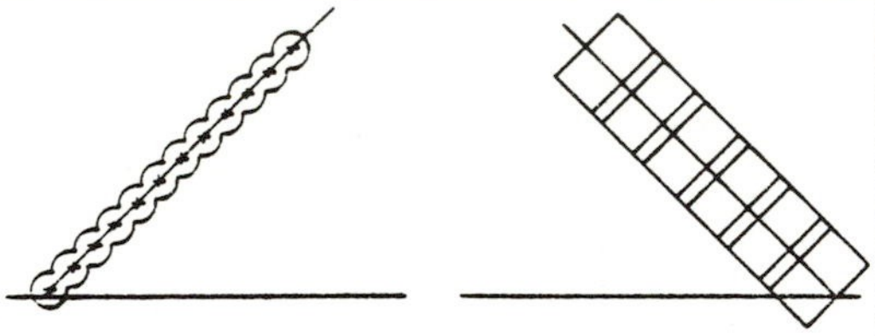
HOLE/w, h, o, $\langle$ANG, $a\rangle$	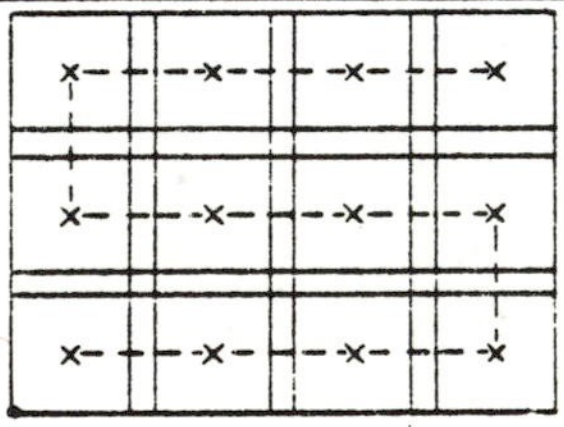
TRIM/w, h, o, $\langle$ANG, $a\rangle$	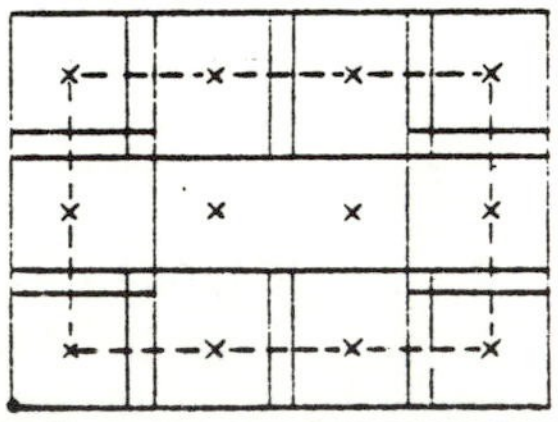

COMPLEX AUTOMATIC PROFILING
ELEMENT

FIGURE 8

Figure 4. Notice that the program language statements are relatively straightforward in meaning and easy to read as contrasted with the manual tape program in Figure 5. Also, the use of absolute (i.e., dimensions taken directly from the print) coordinate, etc. data instead of the incremental data used previously.

```
HUSMAN          09:09EDT      08/31/79

100  OPTION/1,1,1,1,0,0,401
110  PART/HUSMAN DEMO
120  SHAPE/RECT,4.0,2.875,THK,.062,CRS
130  DEFINE/HPAT,1
140  TOOL/02,RND,.875
150  PT/1.375,.625,PRIO,99
160  TOOL/08,RND,.125
170  AT/1.094,1.994$LINE/U,1.25,2,PRIO,98
180  TOOL/05,RECT,.625
190  PT/1.687,1.812,PRIO,97
200  TOOL/11,RECT,.5,3.0
210  PRIO/96
220  AT/-.25,0$PLINE/U,3.375,.1
230  TOOL/20,RECT,.25,1.0
240  PRIO/95
250  AT/2.125,0$PLINE/U,1.5,.1
260  AT/1.5,3.375$PLINE/D,1.5,.1
270  TOOL/19,RECT,1.0,.25
280  PRIO/94
290  AT/0,-.125$PLINE/R,2,.1
300  AT/0,3.5$PLINE/R,1.375,.1
310  END
320  AT/0,0 $ HPAT/1
330  FINI
```

DEMO BRACKET PART PROGRAM

FIGURE 9

After this point, NC program development differs substantially from that of manual NC programming. With computer-assist programming, the user may preview the part drawn to exact scale on a graphic plotter in his office to determine part accuracy. Figure 10 is an example of an NC, computer-assist, graphic part plot. Once a part passes this level of development, the user can be over 99% confident that he has a good part for later use. At this point, a tape could be generated for single part production, or with but a few additional program lines, the single part can be completely propagated over a sheet so as to minimize scrap.

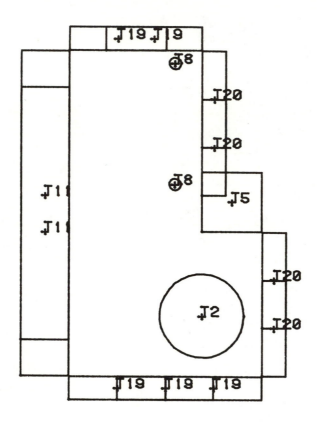

DEMO BRACKET PART
FIGURE 10

Automatic Pattern Layout

Advanced, NC, computer-assist, part program systems offer features not generally found in other NC, computer-assist, part programming systems. These features are also very expensive to replicate manually, but are most desirous because of the impact they have upon man and machine productivity. The ability to lay many parts out on a sheet and guarantee shearing integrity or shaker configuration requires a substantial amount of time, if done manually. SDRC PAL enables the part shown in Figure 10, along with its associated hits, to be propagated across a sheet with the following additional program lines shown below in Figure 11.

```
MAXHMLAY      09:10EDT      08/31/79

100 OPTION/1,1.1,0,1,0,401
110 PART1/HUSMAN
120 STOCK/19.2,36.0,MARGIN,.750,2.0,0,0,THK,.062,CRS
130 WRKHLD/C,8,28
140 LAYOUT/MAX,PART1
150 FINI
```

                    SDRC PAL PROGRAM

                      FIGURE 11

The resulting part can be seen in Figure 6 along with workholder clamp locations. The SDRC Pal System program shown in Figure 11 provides substantial features to minimize programming activity. It allows the programmer to define in the OPTION line the amount of part coordinate data to be printed by the system to aid in debugging the program and inspecting the part. It also enables the part programmer to specify a different object machine for which the tape is to be prepared.

SDRC PAL can automatically call up to ten different parts together and allow the programmer to place them directly on the sheet or automatically place them so as to minimize scrap as well as programming time. It will report material discrepancies if, for example, the large sheet had specified something in the STOCK statement (line 120) of Figure 11 different from the material of the individual piece-parts as specified in the SHAPE statement of Figure 9. This includes material discrepancy in either type or thickness or gage.

Automatic part rotation can be requested by the programmer to better utilize material. The system will report all resulting tooling conflicts by both an error message and a graphic plotter error. The plotter will also verify workholder clamp location as shown in Figure 6 and 16 and repositioner hold-down pad locations when appropriate.

SDRC PAL assembles parts and always guarantees shear line integrity. It can, for example, provide an NC tape for the Wiedemann Optishear System or pin locator holes for shearing.

Many times a part exceeds the useful table area of a machine, thereby, necessitating X-axis repositioning. SDRC PAL provides programmer directed or automatic sheet repositioning. PAL also provides automatic repositioning, where possible, to take hits found in no-hit zones generally associated with workholder clamps, etc.

AUTOMATIC OPTIMIZATION

Many NC sheetmetal shops, after struggling to get this far would gladly call it a day after seeing a complete sheet layout as shown in Figure 6. In fact, some NC, computer-assist, sheetmetal programming systems also might call it a day. An examination of machine table motion shown in Figure 12 for a mirrored, part translation environment, indicates that tool-change time would become unbearable and that hits need to be also scheduled by tool. That is to say, after tool hit priorities are recognized all hits should be collected in tool sequence to minimize tool change time. Tool hit sequencing results in a machine table motion for the Demo Part Bracket Sheet as shown in Figure 13. SDRC PAM offers further optimization by simultaneously considering all the move strategy permutational and combinational alternatives available for the 720 hits and 6 tool changes to be made while processing the sheet. Because of the enormous amount of data to be evaluated, a hard surface, fast disc drive system is required to assure timely processing. SDRC PAM will accomplish this 720 hit task in less than 10 minutes on a minicomputer or in several seconds on G.E. MARK III® timesharing. The end result can be seen in Figure 14. Comparative data is offered below in Figure 15 to evaluate the three reported sheet optimization situations of Figure 12, 13, and 14 respectively. Basically, machine cycle time can be reduced by as much as 46% with SDRC PAM. If, for example, a firm experiences a 70% utilization rate on a machine worth $100/hour and it operates two shifts, this savings might be worth the following:

Hours/Shift	40
x No. of Shifts	2
TOTAL HOURS	80
x Utilization	70%
TOTAL PRODUCTION HRS.	56
x Full Cycle Minimization	46%
TOTAL SAVED HOURS	25.7
x Hourly Rate	$100
NET SAVINGS/WEEK	$2570

It is interesting to note that an additional 18.5% of cycle reduction is possible beyond the conventional tool-sort type of minimization by using the matrix process as employed by SDRC PAM in Figure 14.

JOB DOCUMENTATION

With optimization complete, all that remains is to create a tape for the machine tool control unit and send it along with setup instructions to the shop floor. A Job Setup Data sheet, such as shown in Figure 16, provides the machine operator and/or setup man with sufficient data to expedite job

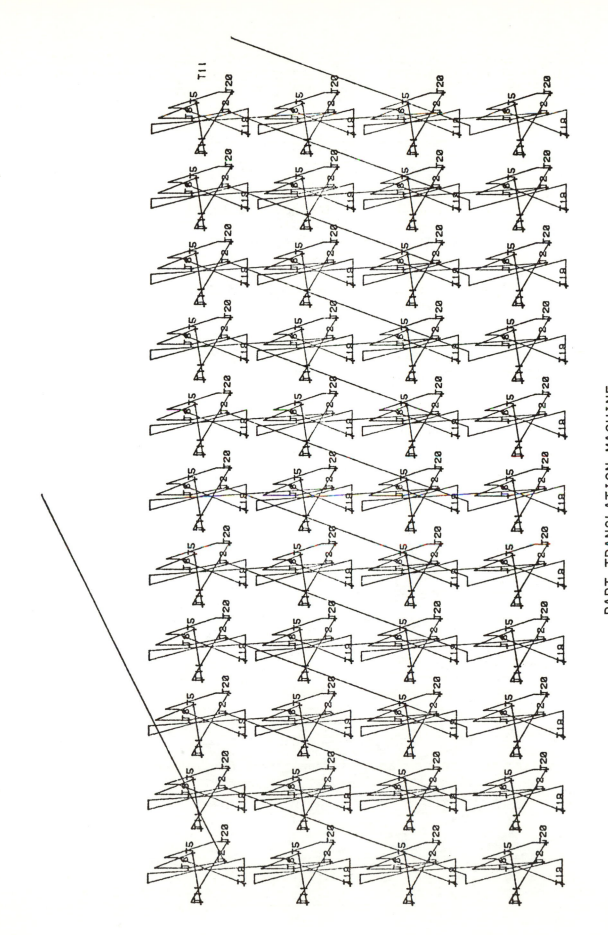

PART TRANSLATION MACHINE
TABLE MOTION
FIGURE 12

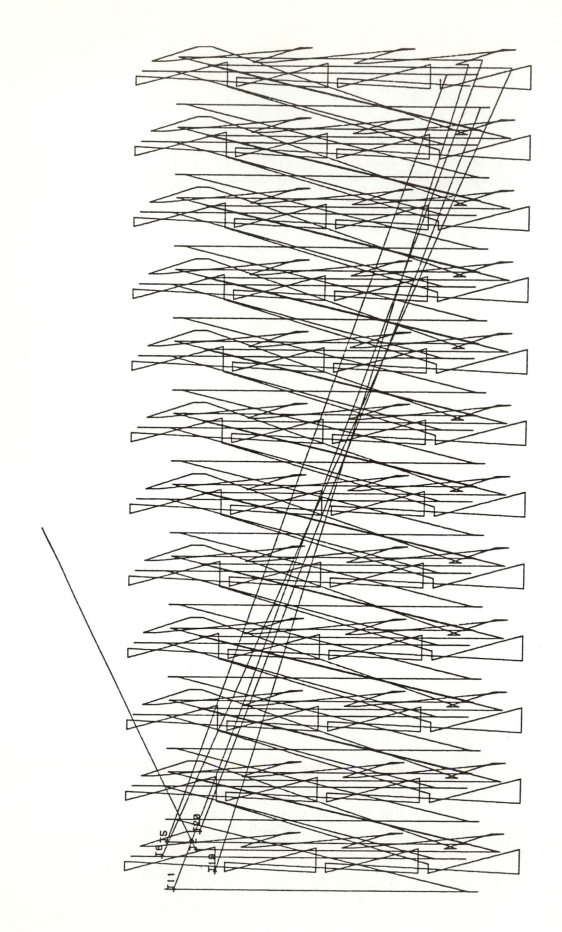

TOOL SHEET OPTIMIZATION MACHINE

TABLE MOTION

FIGURE 13

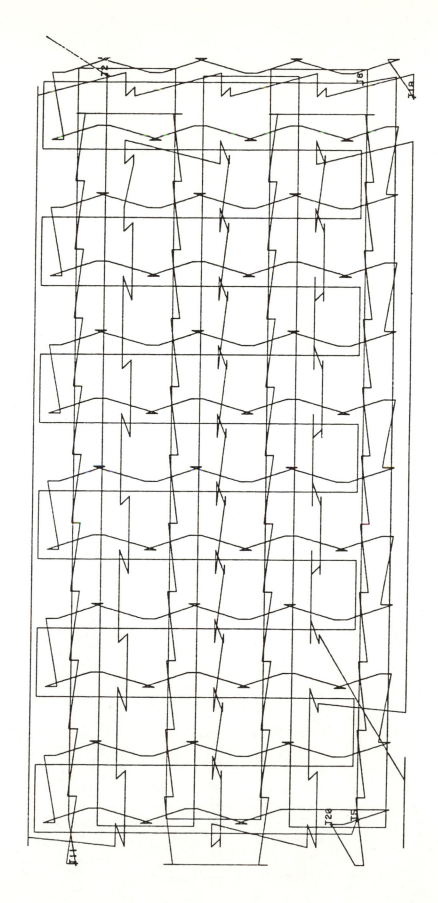

COMPLETE MATRIX OPTIMIZATION
MACHINE TABLE MOTION
FIGURE 14

AREA OF CONCERN	MIRRORED PART TRANSLATION OPTIMIZATION	TRANSLATION & TOOL SORT OPTIMIZATION	TRANSLATED TOOL SORT AND MATRIX OPTIMIZATION
Tool Change Time (Min.)	3.54	.06	.06
720 Hits (Min.)	4.29	5.17	4.20
TOTAL TIME (Min.)	7.83	5.23	4.26
Reduction (Per Cent)	--	33%	46%
TAPE LENGTH (Feet)	91.73	76.70	76.86

OPTIMIZATION ANALYSIS

FIGURE 15

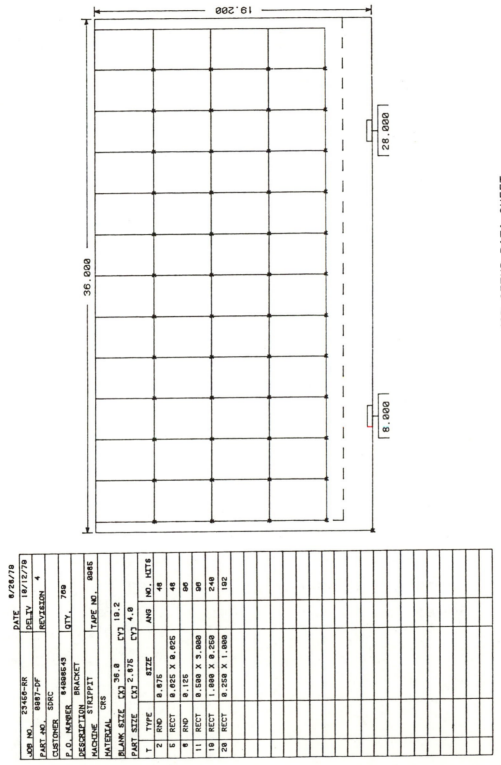

JOB SETUP DATA SHEET
FIGURE 16

JOB NO. 23456-RR		DATE 8/28/78	
PART NO. 0087-DF		DELIV 10/12/78	
CUSTOMER SDRC		REVISION 4	
P.O. NUMBER 84008543		QTY. 760	
DESCRIPTION BRACKET			
MACHINE STRIPPIT		TAPE NO. 0985	
MATERIAL CRS			
BLANK SIZE [X] 36.0		[Y] 19.2	
PART SIZE [X] 2.876		[Y] 4.0	

T	TYPE	SIZE	ANG	NO. HITS
2	RND	0.875		48
5	RECT	0.625 X 0.625		48
8	RND	0.125		96
11	RECT	0.500 X 3.000		96
18	RECT	1.000 X 0.250		240
20	RECT	0.250 X 1.000		192

setup and checkout. The sheet layout shown in Figure 16 could just as readily included all the graphic data found elsewhere in this paper which was displayed by a plotter.

Tabular data found on the left-side of the Job Setup Data sheet relates to tooling setup and manufacturing administration. Additional information regarding job processing and routing is also available to aid as a job router.

GRAPHICS

All through the NC, computer-assist, part programming portion of this paper, a heavy reliance has been placed on graphic validation. Graphic validation allows the NC programmer to visually verify every step he has completed, and be 99% confident of production quality first parts. Graphic validation materially increases productivity of man and machine alike by substantially reducing part proveout time. It allows the part programmer or machine tool manufacturer to synthesize the operation of a particular machine tool without having the machine tool present. Imagine what the concept of competitive machine tool simulation can do for users and builders alike during an evaluation of competing brands of NC machine tool. Or, for that matter, running the processor several times across different machines in order to select the most economic machine for a long run job.

ECONOMICS

The Demo Bracket Part as presented in Figure 6 represents a substantial economic opportunity if NC, computer-assist, part programming is used to create it. Consider the following in order to make a reasonably efficient tape manually:

Program Size (hits)	720	
x Time to Create & Test a Punch Hit (minutes)	2	
TOTAL PROGRAM TIME (hrs.)	24	
x Programmer Cost (per hr.)	$15	
		$360
2 hrs. Machine Proveout Time ($100 per hr.)		200
TOTAL COST TO CREATE PART		$560
Average Cost per Foot of Tape Based on 76.7 Feet		$ 7.30

NC, computer-assist, part programming on the other hand offers the following:

Programmer Time (hrs.)	4	
x Programmer Cost (per hr.)	$15	
PROGRAMMER COST		$ 60
Computer T/S Cost		$ 25
½ Hour Machine Proveout Time ($100 per hr.)		$ 50
TOTAL COST TO CREATE PART		$135
Average Cost per Foot of Tape Based on 76.7 Feet		$ 1.76

With a minicomputer, the processing cost would diminish to $10 and the overall cost to $125 or less than ¼ of the cost for manual tape preparation.

All this translates to the following:

- Increase press work capacity as much as 32%.
- Increase programmer productivity 3 to 6 times.
- Reduce tape preparation cost 3 times.
- Reduce machine tool wear on ball screws, etc.
- Reduce production bottlenecks and work-in-process inventories.
- Increase productivity in terms of value added per manufacturing employees.

TOMORROW

We can only speculate on the future. However, the general direction it appears to be taking is relatively clear because the future is bottom-line (i.e., profit) driven. A manufacturing or jobbing operation is a value-added environment. Material passes through these operations and thereby, becomes more valuable because of machine, assembly, etc. Mr. James Geier, President of Cincinnati Milacron, was cited in the 1979 Winter issue of MSU BUSINESS TOPICS as saying that a part spends only 5% of its manufacturing time on a machine in the average machine shop. Furthermore, only 30% of that time (or 1.5% of the total time) is in the actual machining process.

Clearly, reduction of work-in-process (i.e., WIP) inventory represents a substantial opportunity for improved profitability as we will see later.

Programming & Control

More shops which had programmed manually earlier, are beginning to employ NC, computer-assist, programming via public timesharing initially and later on their own in-house minicomputers. Shops which had owned limited type programming systems are moving up to newer system models in spite of the fact that the older model still functions well because of the improved productivity these new systems bring with them. Computer Numerical Control (i.e., CNC) is becoming increasingly popular because of on-board program tape storage and data editing features. All new NC machine tools are CNC equipped as well as additional NC control units are being field retrofitted with CNC adapter units. In effect, the field retrofit CNC phenomena is a mid-life kicker which effectively lengthens the productive life of an NC machine tool.

Computers

The computer is at the heart of most improvement we see in productivity. This is caused by the computer's very nature (i.e., control). Minicomputers, and now microcomputers, are serving as control unit devices for NC machine tools. Computers are serving as programming systems to generate NC program tapes for NC machine tools. Computers are serving to tie manufacturing and engineering together with CAD/CAE/CAM. This means Computer Aided Design/Computer Aided Engineering/Computer Aided Manufacturing.

Computers will, as suggested, become increasingly involved in every facet of day-to-day job shop operations. The computer aided drafting systems which make part drawings and NC tapes today, will sell for $50,000 within seven (7) years based on the current rate of decline in computer prices. This is due to the technological explosions occurring in the field (e.g., Winchester Storage, bubble memory, 64 bit chips, etc.). The same $50,000 system will simultaneously support designers, programmers, manufacturing managers, controllers and other management from highly centralized design, material, administration and financial data bases. Orders entering a system will be processed by material requirement planning subsystems, process planning systems, shop scheduling systems, etc., with human intervention being required for exception management only.

All this will not happen with the dawning of 1986. Rather it is already beginning to happen. Today, a wide variety of minicomputer systems are being offered for all the various applications mentioned earlier. The difference between today and tomorrow lies in the lack of integration to be found among the disjointed applications available. During the next several years the challenge for the manufacturing systems and software firms will be to complete the integration activity. Data should only be entered into a system once and then be automatically purged when no longer needed.

Strategic Planning

Whether the light you see in your tunnel is real or a train coming your way, depends on how firmly you reign your destiny. Not only will tomorrow's job shop manager/owner be caught up in the technological aspect of his career, but he will also be very much more involved in planning and control. Peter Drucker once said, "The question isn't whether we're doing things right, but increasingly, whether we're doing the right things."

Clearly, tomorrow's winner will be today's planner. The problems associated with the quickening business pace we find ourselves in today, doesn't permit as much latitude in decision making as previously. It's important that today's management plans its work and then works its plan daily. While this session isn't intended to make you strategic planners, it is hoped that with the aid of the next few ideas, you can assimilate all the facts you need to help you determine whether there is light or a train at the other end of the tunnel. The Strategic Planning Model as found in

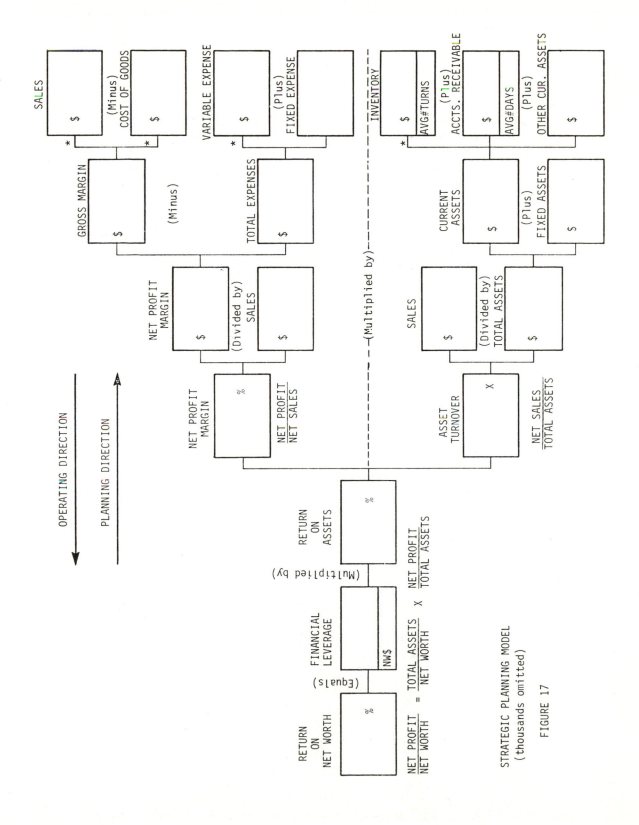

OPERATING DIRECTION

PLANNING DIRECTION

SALES
$ *

(Minus)
COST OF GOODS
$ *

GROSS MARGIN
$

(Minus)

NET PROFIT
MARGIN
$

(Divided by)
SALES
$

NET PROFIT
MARGIN
%
NET PROFIT / NET SALES

VARIABLE EXPENSE
$ *

(Plus)
FIXED EXPENSE
$

TOTAL EXPENSES
$

(Multiplied by)

SALES
$

(Divided by)
TOTAL ASSETS
$

ASSET
TURNOVER
X
NET SALES / TOTAL ASSETS

INVENTORY
$ *
AVG#TURNS

(Plus)
ACCTS. RECEIVABLE
$
AVG#DAYS

(Plus)
OTHER CUR. ASSETS
$

CURRENT
ASSETS
$

(Plus)
FIXED ASSETS
$

RETURN
ON
ASSETS
%
NET PROFIT / TOTAL ASSETS

(Multiplied by)

FINANCIAL
LEVERAGE
NW$
TOTAL ASSETS / NET WORTH

X

(Equals)

RETURN
ON
NET WORTH
%
NET PROFIT / NET WORTH   =   TOTAL ASSETS / NET WORTH

STRATEGIC PLANNING MODEL
(thousands omitted)

FIGURE 17

99

Figure 17 is a means whereby Manufacturing Management can effectively evaluate the impact of its decisions on the future in terms that everyone may understand. When used as a planning device, data is fed by various levels of management and flows from left to right. When used as an operating device, data generated by our daily actions enters on the right side and proceeds to the left side.

A brief example is offered to demonstrate how seemingly small things done by manufacturing can produce gratifying results for the stockholders or owner(s) of a firm. Figure 17a and 17b represents a medium size jobbing firm suffering from among other things excessive work-in-process inventories as cited earlier by Mr. Geier. Presently, this firm realizes 6.7% Return on Net Worth. If these owners could sell their assets at full value, they might be more prudent investing elsewhere just to remain abreast of the present inflation rate and to prevent further asset erosion. However, manufacturing can provide substantial aid because it is typically the largest consumer of expenses and capital in a product-oriented company.

Consider the following opportunities available to our medium size example firm to reduce work-in-process inventory:

● Effective material requirements planning
● Integrated, computer aided design, engineering, process planning and manufacturing
● Effective machine production and maintenance scheduling
● Replacement of existing machine tools with CNC

These actions might reasonably be expected to accomplish the following:

● Increase sales $200,000 (4.2%) by reducing backlogs
● Increase cost of goods sold by $100,000 to support additional sales
● Reduce variable expenses (e.g., labor, tooling, etc.) by $200,000 (7.4%)
● Reduce inventory (i.e., work-in-process and finished goods) by providing shortened manufacturing cycle .4 times or 17%
● Increase accounts receivable because of additional sales

Figure 17b shows the collective value of all these actions to result in a Return of Net Worth of 16.6%. This is a marked improvement over the 6.7% which existed before the change. More importantly, everyone's contribution to the firm's overall profitability can be better understood and their actions guided accordingly.

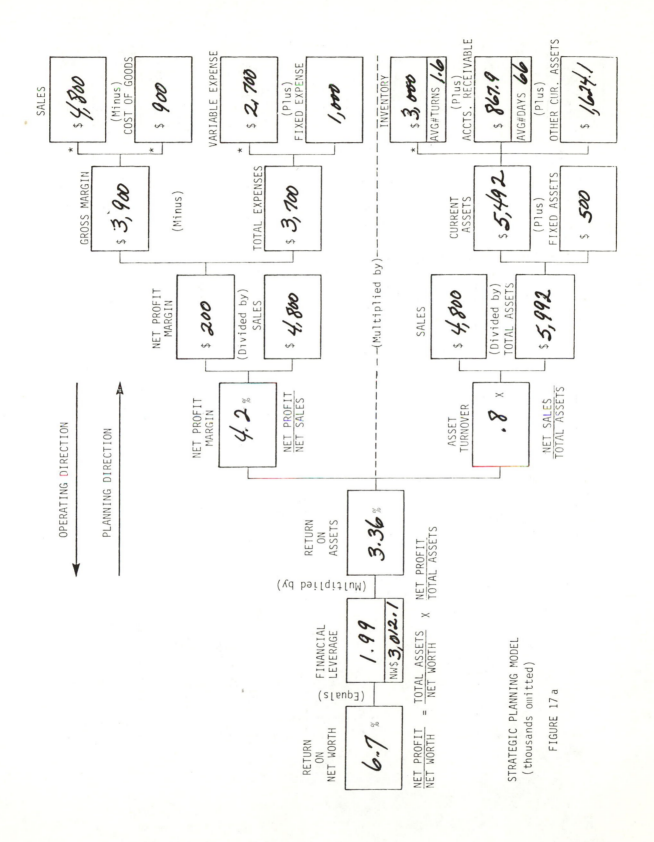

OPERATING DIRECTION

PLANNING DIRECTION

SALES $ 4,800

(Minus) COST OF GOODS $ 900

GROSS MARGIN $ 3,900

(Minus)

VARIABLE EXPENSE $ 2,700

(Plus) FIXED EXPENSE 1,000

TOTAL EXPENSES $ 3,700

NET PROFIT MARGIN $ 200

(Divided by) SALES $ 4,800

NET PROFIT MARGIN 4.2 %

NET PROFIT / NET SALES

INVENTORY $ 3,000

AVG#TURNS 1.6

(Plus) ACCTS. RECEIVABLE $ 867.9

AVG#DAYS 66

(Plus) OTHER CUR. ASSETS $ 1,624.1

CURRENT ASSETS $ 5,492

(Plus) FIXED ASSETS $ 500

SALES $ 4,800

(Divided by) TOTAL ASSETS $ 5,992

ASSET TURNOVER .8 x

NET SALES / TOTAL ASSETS

(Multiplied by)

RETURN ON ASSETS 3.36 %

NET PROFIT / TOTAL ASSETS

(Multiplied by)

FINANCIAL LEVERAGE 1.99

NW$ 3,012.1

TOTAL ASSETS / NET WORTH

(Equals)

RETURN ON NET WORTH 6.7 %

NET PROFIT / NET WORTH = TOTAL ASSETS / NET WORTH X NET PROFIT / TOTAL ASSETS

STRATEGIC PLANNING MODEL
(thousands omitted)

FIGURE 17 a

101

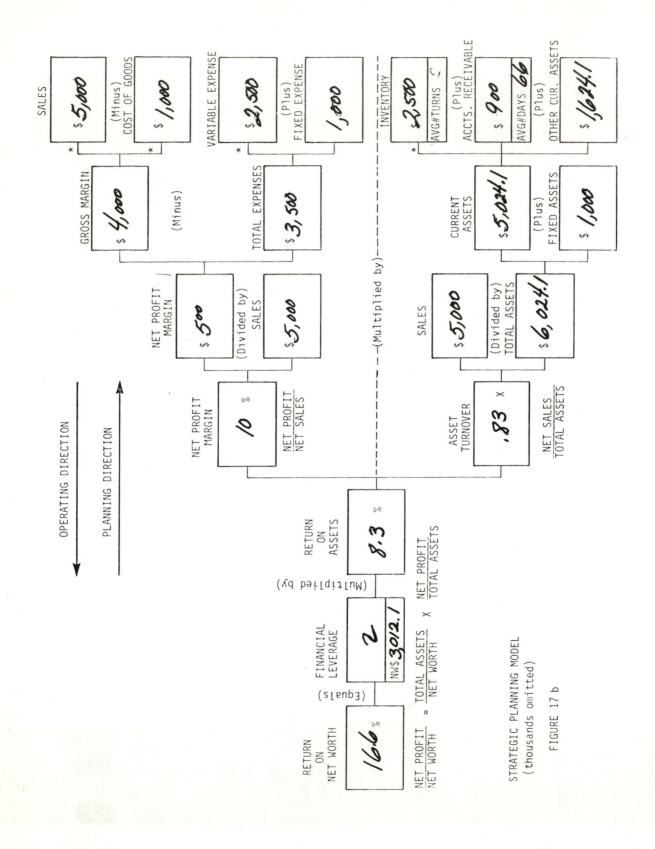

STRATEGIC PLANNING MODEL
(thousands omitted)

FIGURE 17 b

102

CHAPTER 3
SMALL SHOP USES

NC Band Saw Used For Cutting Aluminum Dies

By Arthur R. Meyers
Director of Training
DoAll Company

Today Aluminum Extrusions Dies are manufactured mainly by the
Electrical Discharge Method with carbon electrodes utilized
exclusively. Wire EDM is not used because of the time and
investment necessary. The production of electrodes is basically a hand operation and has kept the cost of the dies high.
Computerized Numerical Control (CNC) can play a large part in
reducing the costs of these dies. Programming directly on the
machine into a memory either with a tape or entry by the operator can reduce substantially the time necessary of laying out
on a template and the machining manually from the pattern.
Using a coated wire band operating at the correct speeds and
feed the electrode can be made in a fraction of the time that
present methods require and with CNC any intricate shape can be
formed. Based on research and industry tests the advantages
and limitations of wire machining in the industry are discussed.

INTRODUCTION

The NC Wire Band Saw for Aluminum Die Making has advantages
over the present method:

- Greater speed of production
- Greater accuracy possible
- Less manpower required
- Cost savings
- Elimination of electrodes in many areas

Most manufacturing of Aluminum Extrusion Dies involve an operator following a template while machining an electrode. This
is a hand operation and requires close tolerances. To achieve
this the worker must proceed with care in producing first the
template and second the electrode. Consequently attention
must be given to simplifying the method to reduce the cost of
production.

A machine to be used must comply to at least three criteria:

1. Economic Criteria: Purchase price in relation to
   production, easy to maintain, reliability, etc.
2. Job related criteria: The machine must be easy to
   operate, programming must be easily accomplished,
   training must be available, machining time must be
   extremely fast, and floor space must be considered.
3. Acceptability Criteria: The machine must be accepted by shop personnel and secondary management.

The wire band saw is related to conventional band sawing in that both are chip removal processes. They utilize an endless band and are governed by the speed of the band and feeding of the work for optimum results. At that point they part company. Conventional band sawing has thousands of small cutting tools, all single point held together by the back of the band that is essentially their tool holder. The greater the back of the band the more rigid the tools are held. Speeds are low and removal of metal is governed by the hardness of the material being cut.

The wire band saw is essentially a grinding operation using extremely fast speeds for metal removal. Hardness of material plays a part but as a whole does not limit the cutting ability of the wire. The wire band is thin and somewhat fragile so extreme caution must be used when machining, hence the NC control. The thinness of the band allows for smaller radii to be cut than conventional band sawing, thereby offering complex shapes to be formed at very high cutting rates.

ALUMINUM EXTRUSION DIE MANUFACTURING REQUIREMENTS

In the present practice of manufacturing Aluminum Extrusion Dies the following procedure is generally used:

1. A template is designed from the blue print, layout is made, usually sawed and finally file finished.

2. An electrode is then made using the template for guidance. The procedure is simple but time consuming. A die filer is rigged so the template can be followed. The graphite is roughed out on the machine to the template shape. It is then taken to another machine where it is finished to shape.

3. The template is then used to layout the back of the die for machining the relief in the die.

4. Finally the shape of the extrusion is burned in by EDM with the electrodes.

Depending upon the complexity of the die the time required for the procedure is anywhere from 3-10 hours and depending upon the number of dies being manufactured the manpower requirements range up to 15 men for electrode manufacturing. Variations of manufacture occur in many shops but manpower requirements and time consumed are essentially the same.

Jobbing shops overcome some of the costs by maintaining lower overhead, however as their business grows they are finding it increasingly difficult to remain competitive. Large companies have tremendous difficulties competing due to their extremely high overheads. They must find better methods to produce these dies. The wire band saw offers some solutions to these problems.

N/C WIRE BAND SAW

The NC Wire Band Saw utilized three types of coated bands for stock removal - Aluminum Oxide, Borazon and Diamond. In addition a spiral saw band can be used. The Borazon band is used for hardened steel and the Diamond band is used for carbides, refractories, glass and other brittle materials. Aluminum Oxide is used for Graphite, Carbon, and mild steels while the spiral band is good for graphite, carbon and soft materials such as many aluminums.

These are endless bands and are joined together using a specially built welder designed for thin wires. (Note Figure 1) Since the wires are coated with nonconductive materials special flux is needed to coat the ends to be joined. The weld is critical for fatigue can occur causing premature band breakage. Welding by the user is mainly for internal machining.

The machine is mounted on a frame capable of driving wheels 26" in diameter. The X and Y table has a total travel of 10" and is driven by two stepper motors with .0005 resolution. The standard controller is a Summit CNC and has both linear and circular interpolation. Any controller can be retrofitted for any special requirements that may be needed. The memory capability is 1025 commands. A tape reader or cassette is available. Controller will accept all standard language which is useful when computor aided programming is used. (Figure 1)

The bands are guided by roller type guides with opposing offset rollers. There are four rollers on both the upper and lower guide posts with class 7 bearings, forced lubricated, mounted in each one. Wheels are mounted on each bearing assembly and covered with a special replaceable plastic tire. The whole assembly is adjustable to line up and guide the wire band. (Figure 2)

The power for the NC Machine is supplied by a D.C. Variable speed drive having a range from 1000 SFM to 5000 SFM. The controls are located on the frame of the machine. Mounting the work to be machined is accomplished through two T slots with clamps. (Figure 3)

MAKING ALUMINUM EXTRUSION DIES WITH NC

The initial operation of any type of Numerical Control is the program needed to command the machine. In the making of electrodes the shape determines the complexity of a program. In addition the number of electrodes required for a job is only a few with the job itself limited to one piece. The economics of programming is of utmost concern to the industry as a whole. Most shapes required are relatively simple. Mostly straight lines with a few contours and radii involved. A typical program will take from an hour to an hour and a half. CNC is the best for it allows easy correction without the cumbersome job of correcting a tape. In addition if a tape is required it can be made right from the controller to a teletype machine for future use. When re-entering the tape with a tape reader the procedure is very fast, entering over 250 commands in less than a minute. For the small shop that has never had NC before, instruction in programming must be supplied. With CNC no special procedure is necessary to program, the number of digits are not important and just a few requirements must be followed.

Machining graphite for electrodes is extremely fast. Whether the material is 1/2" or 2" thick the cutting rate is 1 linear inch per minute. For a two inch piece this represents 60 square inches per hour. For stock removal alone a typical die taking four of five hours under the present method requires only 10 to 15 minutes. The cutting rate is the same for contouring as well as straight cutting. (Note Figures 4, 5 and 6 and the times of cutting.)

The type of graphite represents no obstacle to cutting, whether KK8, KK 10, 12 or 14 or copper graphite the cutting rates are the same. The advantage of the denser and harder graphite is both in finish and life.

Some Aluminum Dies can be made directly with the wire band machine, however there are limitations, such as the smallest radius required and the machining itself. Wire band machining does not lend itself in steel to partial cutting by the wire. Deflection can occur therfore full stock removal by the band is necessary. Where the conditions exist for machining savings can be great.

Other limitations are the thickness of the material being cut. Up to one inch can be machined effectively with cutting rates decreasing as thickness increases. Direct machining of the extrusion die should be done after the reliefs in the back of the die have been machined.

Since this type of die making is all internal the skill of the machine operator welding the blade is important. While it is not difficult welding must be done with care and caution.

There are two sizes of Aluminum Oxide coated bands, .046" and .078". These are nominal sizes and can vary several thousandths. Before programming a sample cut should be made and the kerf checked for size. If the CNC has an offset adjustment, a correction can be made after programming. (Figure 7)

The life of the band while cutting graphite is exceptionally good. Bands have been used for up to 45 hours with no appreciable wear. When cutting steel, wear of the band is a factor. The bands have lasted up to six hours depending upon the thickness of material removed.

When cutting complex shapes in steel, radii and other shapes require dwells built into the program to allow the band to catch up. Graphite does not need the dwells. The controller allows for any time dwell needed, however one to two seconds is usually enough.

CONCLUSION

There is no question about the role the wire band saw can play in this industry. The cutting rates of one linear inch per minute are remarkable when measured against present practice. The accuracy of numerical control is unquestionable. Combined with the machine it offers greater accuracy at greater speeds. Less manpower is required in the die shop. A programmer or two combined with an operator can replace all who presently spend their time making electrodes and templates. The machine can produce both economically. The cost savings in labor, space and time are great and when combined with a direct die production the savings are even greater. Programming requires new skills but with proper training and computor assist where it may be needed, costs of programming can be held to a minimum. NC is not always acceptable to all shops. Fear of programming, loss of a job and skills being replaced are some of the problems new machines face. It remains up to management to alleviate these fears in order to offset reaction in the shop. Personnel can pretty much discourage new equipment if they are not won over to it.

Figure 1 - NC Wire Band Machine with its Controller & Welder

Figure 2 - Guide Assemblies

Figure 3 - Machine Controls

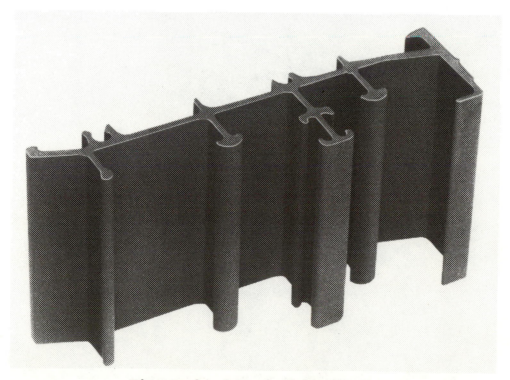

Figure 4 - A Typical Electrode

Figure 5 - A Typical Electrode

Figure 6 - A Typical Electrode

Figure 7 - A portion of an Aluminum Oxide Wire

Improving NC Machine Performance in Job Shops

By Jack Sim
Western Regional Manager
White-Sundstrand Machine Tool Inc.

I purposely selected the phrase JOB SHOP in my title to

A)   Get machining center users to attend this session since they always tell me their manufacturing operation is nothing but a big job shop,

B)   Get a chance to define JOB SHOP,

C)   To emphasize the major point in my paper--that automatic part handling on NC machines applies to job shops.

My definition of job shop can be expressed in five categories:

1. Quality Work

2. Quick Deliveries

3. Low Volume

4. Ingenious Processing

5. No Inventory

If we substitute the word NC for job shop, the definition has amazing similarities. Every category in the above definition is a major benefit in applying NC machines.

The strong move to NC machines in job shops confirms that this comparison in definition is a reality.

If job shops are a natural user of NC machines, how can we improve their performance? The answer is computer managed automation of these machines.

If I were playing a pin ball machine, I probably would get a "TILT" for making that statement. I hope the following explanation eliminates the "TILT".

Since this is a productivity seminar related to NC machines, I'll assume that the audience is basically manufacturing engineers. I will predict that the maximum volume of parts in shops represented at this seminar are under 20,000 per year. This volume applies to parts that have to be made on chip making machines--drill, tap, bore, and mill.

This production volume is not the minority, but it actually represents most manufacturers today. An article in SCIENTIFIC

AMERICAN dated February 1975, estimated that batch-production methods account for 50-75% of national expenditures for manufactured parts. Batch-production was defined by lot sizes from several units to 50 units.

This estimate is confirmed by my own experience in contacting customers who are or will be machining center users. Large corporate giants can be identified with very successful products without mass production numbers in aerospace, aircraft, oil, mining, construction, fluid handling, air conditioning, ship building, generators, missiles, printing, power transmission, packaging, etc.

The traditional method used today to manufacture this category of parts is one operator on each machine to load tools, fixtures, and parts. NC machines position the machine for each operation and tool changers on machining centers take over the tool load functions. All references to these machines will be stand alone.

The computer managed automation of these machines adds computer control of the machines from the office and automatic part load and unload of the machines from a common system. All references to these machines will be manufacturing systems.

In order to compare the benefits, I asked our Manager of Data Processing, Gordon Dirksen, to write a computer program to add up the hours that determine a machine's performance index for one year. The list of items on page 3 determine the categories that we used to identify what a typical machine does during one year. The program is written so we can change one input to determine the change in performance index.

The definition of performance index is:

$$\text{Performance Index (PI)} = \frac{\text{Production hours}}{\text{AVAILABLE hours}} \times 100\%$$

I have purposely not used the word utilization in my presentation since setup hours are given equal recognition to production hours in the NMTBA definition of utilization. The performance index also differs to the utilization value, since it is based on available hours and not 24 hours a day, seven days a week. If a machine can be justified in two shifts, a manager would get a truer picture of the machine's contribution when he compares production hours to available hours.

Page 4 through 7, shows re-runs of the computer program when one value is changed to determine how that value affects the performance index. The Summary on page 7 tabulates all the changes. When the worst performance in a manufacturing system is compared to the best performance with stand alone machines, the difference in dollars produced is $691,047 in a three machine system.

A 42" diameter pallet system on three machines--including computer and installation would cost approximately $760,000. Consequently, a manufacturing system would approach payoff in one year. Naturally, these numbers need specific examination, but I am convinced that our simple computer exercise is close enough to promote an investigation by all job shops using NC. Never before in U.S. history has our industrial ability been challenged in the world market. We may have invented words like production, numerical control, machining centers, computers and automation, but we are overdue in improving our performance index in batch-manufacturing.

A BASE SAMPLE USED FOR COMPARISONS OF THE EFFECT OF CHANGE

HOURS PER YEAR	4000 HOURS
HOURS PER SHIFT	8 HOURS
AVG. PER CENT MACHINE AVAILABILITY	90 PER CENT
AVG. NUMBER OF PARTS PER LOT	10
AVG. NO. OF SETUPS PER PART	3
AVG. CYCLE TIME (MINS.) PER SETUP	60 MINUTES
AVG. SETUP TIME IN HOURS	3 HOURS
AVG. LOAD/UNLOAD TIME IN MINUTES	6 MINUTES
AVG. VALUE OF PART	$1000
INTERRUPTS PER SHIFT	.25
AVG. NO. NEW PROGRAMS PER MONTH PER MACHINE	4
AVG. IDLE MINUTES PER SHIFT	48 MINUTES
NUMBER OF MACHINES IN PROPOSED MFG. SYS.	3

RESULTS OF THE BASE SAMPLE

STAND ALONE SPINDLE PERFORMANCE	=	41.9 PER CENT
VALUE PRODUCED PER STAND ALONE SPINDLE	=	$558,333
MFG. SYS. SPINDLE PERFORMANCE	=	73.5 PER CENT
VALUE PRODUCED PER MFG. SYS. SPINDLE	=	$1,153,208
ADDITIONAL DOLLARS PRODUCED ON A 3 MACHINE MFG. SYS.	=	$1,784,625

```
HOURS PER YEAR-------------------------------------    4000 HOURS
HOURS PER SHIFT------------------------------------       8 HOURS
AVG. PER CENT MACHINE AVAILABILITY-----------           90 PER CENT
AVG. NUMBER OF PARTS PER LOT-----------------          100
AVG. NO. OF SETUPS PER PART------------------            3
AVG. CYCLE TIME (MINS.) PER SETUP-----------            60 MINUTES
AVG. SETUP TIME IN HOURS---------------------            3 HOURS
AVG. LOAD/UNLOAD TIME IN MINUTES------------            6 MINUTES
AVG. VALUE OF PART----------------------------        $1000
INTERRUPTS PER SHIFT--------------------------          .25
AVG. NO. NEW PROGRAMS PER MONTH PER MACHINE-             4
AVG. IDLE MINUTES PER SHIFT------------------           48 MINUTES
NUMBER OF MACHINES IN PROPOSED MFG. SYS.-----            3
```

```
STAND ALONE SPINDLE PERFORMANCE          =       51.9 PER CENT
VALUE PRODUCED PER STAND ALONE SPINDLE   =    $691,740
MFG. SYS. SPINDLE PERFORMANCE            =       75.4 PER CENT
VALUE PRODUCED PER MFG. SYS. SPINDLE     =  $1,183,349
ADDITIONAL DOLLARS PRODUCED ON A         =  $1,474,827
   3 MACHINE MFG. SYS.
```

A CHANGE FROM 10 PARTS PER LOT TO 100 PARTS PER LOT CAUSED A
10% INCREASE IN PERFORMANCE AND A GAIN PER SPINDLE OF
$133,407/YEAR.

                        ****

```
HOURS PER YEAR-------------------------------------    4000 HOURS
HOURS PER SHIFT------------------------------------       8 HOURS
AVG. PER CENT MACHINE AVAILABILITY-----------           90 PER CENT
AVG. NUMBER OF PARTS PER LOT-----------------           10
AVG. NO. OF SETUPS PER PART------------------            2
AVG. CYCLE TIME (MINS.) PER SETUP-----------            60 MINUTES
AVG. SETUP TIME IN HOURS---------------------            3 HOURS
AVG. LOAD/UNLOAD TIME IN MINUTES------------            6 MINUTES
AVG. VALUE OF PART----------------------------        $1000
INTERRUPTS PER SHIFT--------------------------          .25
AVG. NO. NEW PROGRAMS PER MONTH PER MACHINE-             4
AVG. IDLE MINUTES PER SHIFT------------------           48 MINUTES
NUMBER OF MACHINES PROPOSED MFG. SYS.--------            3
```

```
STAND ALONE SPINDLE PERFORMANCE          =       41.9 PER CENT
VALUE PRODUCED PER STAND ALONE SPINDLE   =    $837,500
MFG. SYS. SPINDLE PERFORMANCE            =       73.5 PER CENT
VALUE PRODUCED PER MFG. SYS. SPINDLE     =  $1,729,811
ADDITIONAL DOLLARS PRODUCED ON A         =  $2,676,933
   3 MACHINE MFG. SYS.
```

A CHANGE IN THE NUMBER OF SETUPS PER PART FROM 3 TO 2 DOES NOT
CHANGE PERFORMANCE, BUT MORE PARTS ARE PRODUCED SHOWING A GAIN
PER SPINDLE OF $279,167/YEAR.

```
HOURS PER YEAR--------------------------------------  4000 HOURS
HOURS PER SHIFT-------------------------------------     8 HOURS
AVG. PER CENT MACHINE AVAILABILITY-----------       90 PER CENT
AVG. NUMBER OF PARTS PER LOT----------------        10
AVG. NO. OF SETUPS PER PART-----------------         3
AVG. CYCLE TIME (MINS.) PER SETUP-----------        30 MINUTES
AVG. SETUP TIME IN HOURS--------------------         3 HOURS
AVG. LOAD/UNLOAD TIME IN MINUTES-----------          6 MINUTES
AVG. VALUE OF PART--------------------------     $1000
INTERRUPTS PER SHIFT------------------------       .25
AVG. NO. NEW PROGRAMS PER MONTH PER MACHINE-         4
AVG. IDLE MINUTES PER SHIFT-----------------        48 MINUTES
NUMBER OF MACHINES IN PROPOSED MFG. SYS.----         3

STAND ALONE SPINDLE PERFORMANCE          =        32.6 PER CENT
VALUE PRODUCED PER STAND ALONE SPINDLE   =    $868,519
MFG. SYS. SPINDLE PERFORMANCE            =        70.8 PER CENT
VALUE PRODUCED PER MFG. SYS. SPINDLE     =  $2,222,545
ADDITIONAL DOLLARS PRODUCED ON A         =  $4,062,078
   3 MACHINE MFG. SYS.
```

REDUCING THE CYCLE TIME FROM 60 MINUTES TO 30 MINUTES RESULTED
IN A 9.3% DECREASE IN PERFORMANCE. THE FACT THAT THE VALUE
PRODUCED INCREASED IS BECAUSE MORE PARTS WERE PRODUCED AND THE
PART VALUE WAS LEFT UNCHANGED. OBSERVE THAT THE MANUFACTURING
SYSTEM SPINDLE PERFORMANCE ONLY DROPPED 2.7% WHILE THE VALUE
PRODUCED ALMOST DOUBLED.

****

```
HOURS PER YEAR--------------------------------------  4000 HOURS
HOURS PER SHIFT-------------------------------------     8 HOURS
AVG. PER CENT MACHINE AVAILABILITY-----------       90 PER CENT
AVG. NUMBER OF PARTS PER LOT-----------------       10
AVG. NO. OF SETUPS PER PART------------------        3
AVG. CYCLE TIME (MINS.) PER SETUP-----------        60 MINUTES
AVG. SETUP TIME IN HOURS--------------------         3 HOURS
AVG. LOAD/UNLOAD TIME IN MINUTES-----------          6 MINUTES
AVG. VALUE OF PART--------------------------     $1000
INTERRUPTS PER SHIFT------------------------      .125
AVG. NO. NEW PROGRAMS PER MONTH PER MACHINE-         4
AVG. IDLE MINUTES PER SHIFT-----------------        48 MINUTES
NUMBER OF MACHINES IN PROPOSED MFG. SYS.----         3

STAND ALONE SPINDLE PERFORMANCE          =        45.2 PER CENT
VALUE PRODUCED PER STAND ALONE SPINDLE   =    $602,976
MFG. SYS. SPINDLE PERFORMANCE            =        73.5 PER CENT
VALUE PRODUCED PER MFG. SYS. SPINDLE     =  $1,153,208
ADDITIONAL DOLLARS PRODUCED ON A         =  $1,650,696
   3 MACHINE MFG. SYS.
```

A REDUCTION OF THE NUMBER OF INTERRUPTS FROM 1 EVERY 4 SHIFTS
TO 1 EVERY 8 SHIFTS CAUSED A 3.3% INCREASE IN PERFORMANCE AND
A GAIN OF $44,643.

```
HOURS PER YEAR--------------------------------  4000 HOURS
HOURS PER SHIFT------------------------------     8 HOURS
AVG. PER CENT MACHINE AVAILABILITY----------     90 PER CENT
AVG. NUMBER OF PARTS PER LOT----------------     10
AVG. NO. OF SETUPS PER PART-----------------      3
AVG. CYCLE TIME (MINS.) PER SETUP-----------     60 MINUTES
AVG. SETUP TIME IN HOURS--------------------      3 HOURS
AVG. LOAD/UNLOAD TIME IN MINUTES------------      6 MINUTES
AVG. VALUE OF PART--------------------------  $1000
INTERRUPTS PER SHIFT------------------------    .25
AVG. NO. NEW PROGRAMS PER MONTH PER MACHINE-      8
AVG. IDLE MINUTES PER SHIFT-----------------     48 MINUTES
NUMBER OF MACHINES IN PROPOSED MFG. SYS.----      3

STAND ALONE SPINDLE PERFORMANCE       =       33.3 PER CENT
VALUE PRODUCED PER STAND ALONE SPINDLE =    $444,048
MFG. SYS. SPINDLE PERFORMANCE         =       70.1 PER CENT
VALUE PRODUCED PER MFG. SYS. SPINDLE  =  $1,098,868
ADDITIONAL DOLLARS PRODUCED ON A      =  $1,964,460
   3 MACHINE MFG. SYS.
```

AN INCREASE IN THE NUMBER OF NEW PROGRAMS PER MONTH FROM 4 TO
8 CAUSED AN 8.6 DROP IN PERFORMANCE AND A DECREASED PRODUCT
VALUE OF $114,285/YEAR.

****

```
HOURS PER YEAR--------------------------------  4000 HOURS
HOURS PER SHIFT------------------------------     8 HOURS
AVG. PER CENT MACHINE AVAILABILITY----------     90 PER CENT
AVG. NUMBER OF PARTS PER LOT----------------     10
AVG. NO. OF SETUPS PER PART-----------------      3
AVG. CYCLE TIME (MINS.) PER SETUP-----------     60 MINUTES
AVG. SETUP TIME IN HOURS--------------------      3 HOURS
AVG. LOAD/UNLOAD TIME IN MINUTES------------      6 MINUTES
AVG. VALUE OF PART--------------------------  $1000
INTERRUPTS PER SHIFT------------------------    .25
AVG. NO. NEW PROGRAMS PER MONTH PER MACHINE-      1
AVG. IDLE MINUTES PER SHIFT-----------------     48 MINUTES
NUMBER OF MACHINES IN PROPOSED MFG. SYS.----      3

STAND ALONE SPINDLE PERFORMANCE       =       48.3 PER CENT
VALUE PRODUCED PER STAND ALONE SPINDLE =    $644,048
MFG. SPINDLE PERFORMANCE              =       76.1 PER CENT
VALUE PRODUCED PER MFG. SYS. SPINDLE  =  $1,193,962
ADDITIONAL DOLLARS PRODUCED ON A      =  $1,649,742
   3 MACHINE MFG. SYS.
```

A DECREASE IN THE NUMBER OF NEW PROGRAMS PER MONTH FROM 4 TO
1 CAUSED AN INCREASE OF 6.4% IN PERFORMANCE AND AN INCREASE
IN VALUE PRODUCED OF $85,715.

```
HOURS PER YEAR----------------------------------  4000 HOURS
HOURS PER SHIFT---------------------------------     8 HOURS
AVG. PER CENT MACHINE AVAILABILITY----------        90 PER CENT
AVG. NUMBER OF PARTS PER LOT----------------         10
AVG. NO. OF SETUPS PER PART-----------------          3
AVG. CYCLE TIME (MINS.) PER SETUP-----------         60 MINUTES
AVG. SETUP TIME IN HOURS--------------------          3 HOURS
AVG. LOAD/UNLOAD TIME IN MINUTES------------          6 MINUTES
AVG. VALUE OF PART--------------------------     $1000
INTERRUPTS PER SHIFT------------------------       .25
AVG. NO. NEW PROGRAMS PER MONTH PER MACHINE-          4
AVG. IDLE MINUTES PER SHIFT-----------------         24 MINUTES
NUMBER OF MACHINES IN PROPOSED MFG. SYS.----          3

STAND ALONE SPINDLE PERFORMANCE         =          45.4 PER CENT
VALUE PRODUCED PER STAND ALONE SPINDLE  =      $605,952
MFG. SYS. SPINDLE PERFORMANCE           =          73.5 PER CENT
VALUE PRODUCED PER MFG. SYS. SPINDLE    =    $1,153,208
ADDITIONAL DOLLARS PRODUCED ON A        =    $1,641,768
   3 MACHINE MFG. SYS.
```

REDUCING IDLE TIME PER SHIFT FROM 48 MINS. TO 24 MINS. CAUSED
AN INCREASE IN SPINDLE PERFORMANCE OF 3.5% AND AN INCREASE IN
VALUE PRODUCED OF $47,619/YEAR.

SUMMARY

THE MANUFACTURING SYSTEM SPINDLE PERFORMANCE AND VALUE PRO-
DUCED ARE NOT GREATLY AFFECTED BY CHANGE.

        MIN.   70.1%   MAX.   76.1%

RANGE OF CHANGE IN THE PRECEDING EXAMPLE.

	PERFORMANCE	ANNUAL VALUE PRODUCED PER SPINDLE
STAND ALONE	32.6%	$444,048 MIN.
	51.9%	868,519 MAX.
MANUFACTURING	70.1%	1,098,868 MIN.
SYSTEM	76.1%	2,222,545 MAX.

Reprinted from: Tooling & Production, April 1978

Retirement hobby turns into CNC job shop

When Gus Wittick said goodbye to the metal fabrication business seven years ago and sought a life of retirement ease in Orlando, FL, he had little reason to think that a few years later he would be guiding a modern job shop through a period of rapid growth and expansion.

But that's what happened. Started as a hobby to occupy the time of a man who was too energetic to sit around relaxing, Wittick Metal Products Inc has grown from a $15,000 investment in some used machine tools to the point where it is approaching $1,000,000 in volume. The work force has grown to 25 employees, property was acquired recently to nearly double the existing 20,000 sq ft of plant space, and Wittick has become the owner of a Wiedematic Mach II CNC W-2040 turret punch press.

Gus credits much of the firm's growth to his lawyer son, Gary, who joined the firm four years ago when there were only six employees. Within a year, the firm had acquired its first piece of NC equipment—an older tape-controlled punch press. Gary pushed for the acquisition of the new CNC turret punch press and arranged for leasing it—a radical departure for Gus who had always paid cash for equipment in the past or simply done without.

What Gus didn't know when he started his hobby-business was the unusual nature of the competition. One form consisted of veteran metal fabricators like himself who moved to Florida in retirement and set up small job shops to keep active, with income as a secondary factor. This type of competition is tough because overhead is low and pricing can be highly flexible. On the other hand, there is rarely enough incentive to reinvest profits in new equipment and expansion.

The other type of competition was tougher to meet. This consisted of salesmen and owners of large job shops in Chicago, Milwaukee, New Jersey, and New York who would seek Florida business at rock-bottom prices simply as a means of writing off the expense of a winter vacation. This may have been one reason why it took so long to develop a competitive metal fabrication industry based in Florida.

Sheet metal business

Wittick works primarily with aluminum and mild steel purchased directly from mills in sheet form. The aluminum is supplied in 48" x 144" sheets and the mild steel in 48" x 120" sheets. Sheet thicknesses range from 0.032" to 0.250" in aluminum and 0.035" to 0.187" for mild steel.

As a rule, Wittick prefers to avoid orders for parts larger than 48" for two reasons. Their press brakes cannot handle larger sheet sizes and this is the largest size that can be handled conveniently by one worker without need for a helper.

For holemaking and stamping, Wittick maintains a substantial capacity in high tonnage punch presses with permanent tooling in addition to its older NC punch press and the new CNC turret punch press. But the use of such tooling is steadily declining and with the acquisition of the new Mach II machine, the firm is considering the disposal of some of its older punch presses.

Until recently, Wittick says, if an order for 5000 pcs was received, they would suggest permanent tooling to the customer. Now it is common practice to suggest NC tape-controlled punching for part volumes up to 30,000 pcs.

Furthermore, Wittick has adopted a policy that any part that will fit into the 4" die clearance of the Mach II will be stamped out on that machine.

Higher productivity

The reason for this is the demonstrated higher productivity of the CNC turret punch press. The Wiedematic Mach II CNC W-2040 turret punch press is a 22-ton machine with the capability to accept sheets up to 40" wide. Operated by computer numerical control rather than the hard-wired controls of earlier presses, it is approximately twice as fast as older comparable Wiedemann machines. The hit rate on 1" moves is 190 hits/min and on 10" moves the machine can deliver 90/min. Table travel speed is 2000 ipm.

By contrast, Wittick's older NC machine's hit rate is only 120 hits/min maximum and its table travel speed is much slower.

The greatly increased speed of punching is measured in another way that is important. The older NC press used to operate on a 10-hr shift; the new W-2040 Mach II machine does the same work in only 5 hr, leaving 50 percent reserve

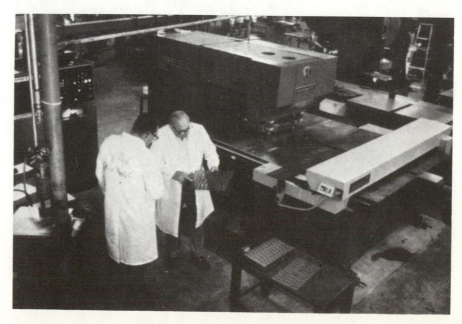

Gus Wittick (right) and MACH II operator inspect part punched by computer numerical control. New turret punch press has 22-ton capacity, takes 40"-wide sheet, and has taken over much of the holemaking and stamping work since its installation.

capacity. One operator is assigned to the machine permanently.

But the machine speed is only part of the reason for its high productivity. Permanent tooling is expensive, takes time to produce and requires tedious setup time. Tapes for the Wiedematic can be produced from the blueprint to the machine in around 4 to 6 hr. What's more, using the 24 tools already in the turret, different parts can be produced with only a few minutes' time required for changing tapes. Repetitive accuracy is very high and inspection requirements after the first part is made are minimal.

To get maximum productivity from the Mach II, Wittick runs multiple parts on single blanks in 24" or 48" lengths whenever possible. A typical example is a small steel bracket which is laid out as 78 pcs on a 24" x 16" blank. Using three tools in the turret for a total of 390 hits, the 78 parts are produced in 2.5 min or 1872/hr. In a standard run of 6000 to 7000 pieces, the entire run can be produced in less than the 20 hr it would take to make the permanent tooling.

This capability is very important because experience indicates that many customers are stocking at lower inventory levels than before and are actually deferring deliveries as a cost reduction measure. To Wittick, accommodating this trend is a competitive advantage gained by the use of the CNC, and further reason to minimize dependence upon punch presses using permanent tooling.

CNC offers another benefit apart from machine productivity. This is in the speed and ease of programming the tapes. All calculations are done on the blueprint and then typed on a tape-making device. With the geometric input option, programmers do not have to manually calculate punching patterns such as arcs, bolt holes, circles or cutouts; these are programmed through the use of simple geometric statements in Wiedepoint, a special machine tool computer language.

The benefit of this system is easily demonstrated. A CNC tape containing 390 hits is 6" long. A similar part programmed for the older NC machine would have required a tape 180" long. In a 2-ft CNC tape, they can program up to 7000 hits. Overall, the programming time for the new machine is 80 percent less than for the older machine.

The Mach II turret punch press was the first piece of equipment ever leased by Wittick. According to Gary Wittick, leasing was considered as a means of conserving capital for planned expansion. The firm found difficulty, however, in arranging favorable terms with local banks which were not sufficiently experienced in financing capital equipment. Eventually, he secured a lease from Warner & Swasey Financial Corp. Besides a special NC lease provision which deferred payments for six months while a tape library was built up, Wittick found that the cost over the life of the plan was lowest of any they investigated. The machine costs the company $5660 quarterly, which, of course, is a write-off against income.

Wittick has a tip for other purchasers of similar machines when their company is located in an area not known for heavy industry. He took advantage of a little-known manufacturer's installation kit which sells for $990 as an option. Consisting of special tapered and predrilled metal squaring plates along with heavy-duty rigging hooks, the kit enabled Wittick to complete installation of the machine in less than three days.

■

CNC control panel has unusual English language display which tells machine operator how to correct a mistake or malfunction. More than 40 such troubleshooting messages are stored in the machine's memory. This feature greatly reduces machine downtime.

This small steel bracket is punched 78 pcs at a time from a 24" x 16" blank. Using the three tools shown and a total of 390 hits, Wittick produces 78 pcs in 2.5 min.

Reprinted from Cutting Tool Engineering, January 1979

8 Tips for the Small Shop N/C User

By George L. Holdridge
Marketing Manager
Moog Inc., Hydra-Point Division

IF YOU are a small shop owner, here are eight suggestions geared to help you get the most out of your N/C purchase — and out of the machine builder, or his representative:

1. Before you make your decision on which N/C machine to purchase, compare cost per part for the work you are now doing on any N/C machine you're considering vs. the cost to do it now with the equipment in your plant. You can calculate the Return on Investment point by dividing the cost of each N/C machine by the annual savings.

2. Compare the customer support materials and programs for each prospective supplier: operator training, maintenance training, programmer training and manuals. Also find out if these programs — both operator-programmer and maintenance — are included in the machining center price. Know what you're paying for — and the proximity of a serviceman and spare parts.

3. Ascertain what utilities and facilities provisions must be made and make them before the machine arrives. If you don't, you could be paying for weeks on an idle machine awaiting installation of a power line or the like. But if you're ready, often an N/C machine can be making parts-and-money for you soon after it arrives.

4. Plan the work, have the programs done, the parts on hand and the operators ready to put on the machine. Orient your people to N/C beforehand if it's your first machine. Select your operator, send him to the manufacturer's school. Have the place primed.

5. Follow the instructions. Often the same traits of independence that motivated the small shop owner to go into business in the first place are the ones that get him into trouble with N/C equipment. The "Just deliver it, I'll make it work" attitude is self defeating. This philosophy might work if you regard yourself as a super shop hand. But if you regard yourself as a businessman and manager, you'll take to heart the advice, suggestions, manuals and materials provided along with the hardware.

6. Send your maintenance man to the machine maintenance school. Most N/C machine makers have such sessions. It's to their benefit as well as yours. Be sure the maintenance procedures are followed faithfully.

7. Call the factory if you have a service problem. Sales representatives aren't really geared to help. They'll only refer you to the home factory anyway — as they should. Sometimes sharp service guys at the machine builder's factory can diagnose and suggest fixes over the phone. It helps the factory to know what things are going wrong and how frequently.

8. The smaller you are, the more important it is to follow the manufacturer's procedures to the letter. Treat the machine right and it will treat you right.

● ● ●

CHAPTER 4

TOOLING

Numerical Control—A Toolmaking Environment

By Edmond P. Walsh
Manager, Equipment Planning
IBM Corporation

Less than thirty years ago, N/C Research and Development began and has progressed rapidly throughout the years. The "Machining Center" of some fifteen years ago was another stimulus to the N/C Machine Tool, and customer exposure, as well as magazine publicity, continued. Today in this country, history reveals the overall acceptance by Manufacturing Engineers, Management, and Company Executives of this method of automated fabrication.

INITIAL INSTALLATION

Back in 1966, IBM in Essex Junction, Vermont, began its N/C awareness and growth. Equipment justification was based on vendor build comparisons, manufacturing lead times, conventional part costs, and improvements in personal productivity. All of these considerations are certainly valid, even today, in our Manufacturing environment. Machining capabilities during this time were somewhat restrictive by today's standards. All Machine Tools were point-to-point with most having only two axis tape control. Programming was manual and cutting tools were somewhat limited by a lack of forecasting knowledge. Work flow was also somewhat of a problem, due to the learning curve of the Tool Planner in his identification of appropriate N/C work. In a short eleven months, N/C had made itself known by timely deliveries, quality parts, and rapid turn around of customer requirements.

TOOLMAKING CONCEPT

The size of the Essex Junction plant and the nature of its manufacturing mission, made it apparent that production parts would not be the substaining force for Machining. There was, however, an ever-increasing demand for Toolmaking of prototypes, low volume fixturing, and process equipment details. It was apparent that certain limitations in programming and Machine Tool capabilities existed and Manufacturing Engineering set about identifying and solving these problem areas.

In order to perform most toolmaking tasks, it was obvious that a turning capability, continuous path machining and precision hole locating were all necessary. With this in mind, Engineering investigated and procured three tools with the capabilities described. (See figures A, B, C).

Figure A
3 - AXIS CONTINUOUS PATH
    MILLING MACHINE

Figure B

AN EXTREMELY ACCURATE PIECE OF EQUIP-
MENT CAN HOLD A TOLERANCE OF PLUS/
MINUS .000095

HAS THE CAPABILITY TO PERFORM "CLOSE"
LIGHT MILLING

Figure C

THIS (8) TOOL TURRET
HAS FULL CONTOURING
CAPABILITIES, AUTO-
MATIC BAR STOCK FEED
AND TOOL OFFSETS IN X
& Z FOR ALL (8) TOOLS.
TOLERANCE CAN BE HELD
TO PLUS/MINUS .001

DEVELOPMENT OF PROGRAMMING

To complement the equipment purchased, efforts began to improve individual programming expertise. The decision was made to explore a universal computer language capable of supporting a tooling environment. APT was selected over less flexible languages and training began for all programmers. We now find ourselves using APT, along with individual machine post-processors, in machining intricate details and generating three axis contours.

This has allowed a normal progression into Sculpturing and TABCYCLE forms required for dies, molds, and tooling. We anticipate DNC installation in the near future with the 5275 IBM Terminal System attached to the three axis contouring millers and lathes. We are looking forward to the application of "On Line" capabilities not yet experienced.

Figure D

AUTOMATIC TOOL CHANGER WITH 4
POSITION INDEX TABLE - 3 -
AXIS MOTION

Figure E

VERTICAL DRILLING CAP-
ABILTY WITH LARGE X-Y
RANGE (54" x 22")

Figures D & E round out the versatile machining capabilities of the Toolmaking center. Many times, two or more N/C machines will be used in completing a detail, as well as other machine tools.

Figure F below, is a prime example of a complete tool utilizing these concepts.

Figure Fa

DETAILS UNASSEMBLED

Figure F

FIXTURE READY FOR USE

Equipment Performance

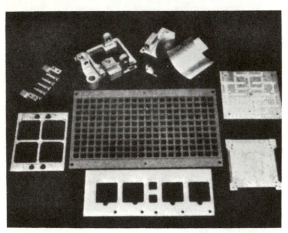

Figure G

VARIOUS PARTS MACHINED.
MATERIALS USED - ALUMINUM,CAST
ALUMINUM, STEEL AND GLASS LAM-
INATED PLASTIC.

TOLERANCES ARE FROM PLUS/MINUS
.002 TO PLUS/MINUS .005.

Many examples exist to highlight the savings realized from this type of operation, and some of these will be presented with the data and background peculiar to the part.

Figure G, contains the first example. The tray in the center of the photo was first machined utilizing conventional machine tools in the Machine Shop. After this, the part was placed in the N/C Shop on a vertical machine and finally worked its way to the Machining Center. The reason for this step progression was the quantity requested by the customer. The basic raw material is G-10 (Glass Impregnated Epoxy) and requires carbide tooling. The data for machines is as follows:

Conventional Machining: 48 Hours/Part

Vertical N/C Machining: 14 Hours/Part

Machining Center: 2.8 Hours/Part

Lot size is generally 25 pieces every three months.

Industrial engineering also conducted an economic evaluation as to the feasability and cost of molding this part - N/C won the study!

Contouring and Sculpturing

Figure H   Figure I

PARABOLIC CONTOURING PROGRAMMED   RIGHT & LEFT, MALE & FE-
$Y^2 + Z^2 = 4\ PX$ MALE - ALL ON THE SAME
                                    TAPE. THIS SCULPTURED
                                    SURFACE WAS NICKNAMED
                                    "THE FLYING CARPET"

The figures above share a common element - that of computer assist programming. Writing an APT manuscript that may take one half of a day, can realistically save many man months of calculations, trial and error.

Figure H was attempted with conventional machine tools and after many attempts and many weeks, IBM, Essex Junction, was contacted to consider the effort. Once the APT manuscript was written, a wooden block was used to prove the contour. Since this first part was machined, many engineering changes have been implemented. All of these changes were over the telephone from a sister plant nearly 1,000 miles away.

Many such situations have occured in the past few years and data has even been transmitted to us over the telephone lines as a test of a program originating at IBM's Scientific Center in Los Angeles.

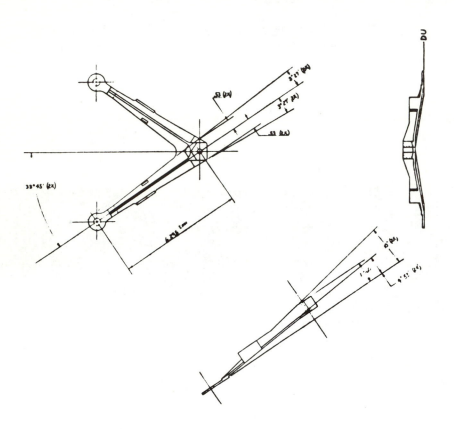

Figure J

Thin Wall Machining

Toolmaking with N/C has the same problems experienced in the Machine Shop. Although the programming of figure J was effectively accomplished using APT, the big problem was support in machining. The thin webs and close tolerancing, contributed to this and the programmer showed innovativeness in the holding fixture design and clamping positions.

The machining cycle involves two position changes and requires less than four hours. Another significant event is that a major engineering change was made on the part and required only four hours of re-programming.

Turn Around Time

Unfortunately, sometimes a disaster is the vehicle that finalizes equipment justification and substitutes the dollars and cents of objects. On January 22, 1970, a blaze leveled the Thermal Wire Electronics Corporation in South Hero, Vermont.

This firm was contracted as a supplier to assemble transformer read - only storage units and wire contact relays for IBM, Essex Junction. Six days after the tragic fire, operations resumed, with some fixturing, at a leased building in Northern Vermont.

An immediate need existed for replacement fixturing and Tooling and the Numerical Control Area met the challenge. They provided more than 60 fixtures representing some 6,000 spindle hours - all the while maintaining an output to previous machining commitments.

Unfortunately, a situation of this type is sometimes the only way to convince company executives of the value of Numerical Control. Fortunately, management was astute in our case and recognized, years before, the assets of Tape Control.

Numerical Control Productivity

In an attempt to quantify our N/C experience during the past nine years, the following rational was pursued:

1.	Select a representative sample of jobs built during the years in question.

2.	Compare N/C hours against conventional build hours.

3.	As insurance to maintain the "Middle of the Road" ratios beyond 20 to 1 were considered. Many such jobs were done during the 1971, 1973 time frame, some of which that could not have been done conventionally.

Productivity Ratio Formula

$$P = H_c/H_{n/c}$$

Where

P = Productivity Ratio

H_c = Conventional Hours

$H_{n/c}$ = Numerical Control Hours

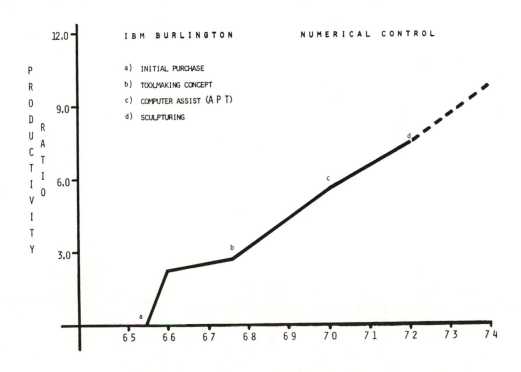

Figure K   NUMERICAL CONTROL PRODUCTIVITY GRAPH

NUMERICAL CONTROL EQUIPMENT

QUANTITY	DESCRIPTION	SIZE	TYPE
2	3 Axis Miller	30"x 48"x 15"	3-Axis Contouring Interpolation, Mirror Image
1	Small Bar, Chucker Lathe	10" Swing 1" Bar	2-Axis Contouring 8 Tool Turret
1	Jig Bore	24"x 14"	2-Axis - Accuracy Plus/Minus .000095
3	3 Axis Miller	24"x 14"x '4"	3-Axis, 4 Position Index Table and Tool Changer 15 Tools
3	Drilling Machine	54"x 22"	2-Axis Drilling and Milling
1	Large Bar Chucker Lathe	16" Swing 3" Bar	2-Axis Contouring 12 Tool Position
1	Milling, Drilling Machine	40"x 20"	2-Axis Contouring Mirror Image
1	High Speed Small Hole Drill	12"x 12"	2-Axis Accuracy Plus/Minus .0005

ATTACHMENT 1

SUPPORT EQUIPMENT

A. Comparator For Lathe Tool Setting.

B. Presetting Tooling Setup For Milling Machines.

C. Quality Assurance Check With The Use Of A Precision
    X, Y, Z, Coordinate Measuring Instrument.

D. Latest Checking Gages Used.

E. Plotter For BCD Format Tape.

F. Regen And Verification Of Tapes.

G. Flexowriter.

H. System 7 Computer, Attached To A Precision X, Y, Z,
    Coordinate Measuring Instrument.

BACK-UP EQUIPMENT

1   Gap Lathe

1   Surface Grinder

1   Hand Milling Machine

1   Band Saw

Assorted Drill Presses

1   Small Chucking Lathe

ATTACHMENT 2

136

Interchangeable Tooling For Machining Centers

By George G. Barkley
Kennametal Incorporated

This paper will describe the development of a new boring system with radial and axial adjustments, for N/C machining centers. It discusses the interchangeability of the tooling system between existing and future machining centers. It will also cover design features and strength evaluation of the adjustable axial connection in a rotating boring bar.

INTRODUCTION

N/C machining centers designed for drilling, reaming, tapping, boring and milling in small to medium production lot sizes, have come of age in the past two decades. They encourage the use of many aids to increase productivity and to assure time-saving repeatability and efficiency. Among the productivity-improving techniques are reduced cycle time, simple tape changes, computer assistance, sufficient tool capacity for varied operations, automatic tool changers, and dual setups or pallet loading for different parts.

Machining centers provide many benefits, all designed to keep that spindle turning and producing. With the average hourly cost of operation pushing well past $25, users cannot afford to idle machines worth $40,000 to more than $400,000. Yet, machining centers present problems. Without sufficient back-up tooling, tool failures ultimately will stop a machine cold, and forced inactivity normally yields significant losses. Many times when machines go down, "Murphy's law" comes into play and normal tool changes can take too long. The dilemma is sharply aggrevated by the lack of interchangeability among tool adaptors supplied by some 25 different builders. In many shops, this variety of separate integral boring bar systems has led to unbearable tool inventories. Innovation was needed to develop optimum machining center adaptor configurations.

Background

To help control tool inventory and random downtime, new tooling systems have been evolved. One of the newest is by Kennametal Inc. of Latrobe, Pennsylvania.

The system gives users total interchangeability of tools (Fig. 1) from one machine to another, in case of tool failure or when back-up tooling proves inadequate. This gives the flexibility needed to reduce unplanned downtime and to improve productivity. It also allows users to buy new and different machining centers without obsoleting existing tool inventories. Such interchangeability can afford obvious productivity improvements.

The new system is designed around the proposed ANSI standard V-flange (Fig. 2) that offers another way to interchange machining center spindle adaptors. This highly desirable proposal is in the hands of the ANSI TC-45 committee and is likely to become official within a year. Kennametal's adaptors are available in 40 mm, 45 mm and 50 mm taper sizes, plus other sizes to fit virtually any machining center. These adaptors accept a variety of bars which slide over a common pilot.

This unusual tooling concept, using axial and radial boring adjustments, permits presetting bars in central tool cribs or at work stations. This helps reduce downtime caused by parts changes.

Programming is simplified by radial/axial adjustments on the bars for predetermined bore diameters and common set lengths. Since each length of bar need not be checked and entered onto the tape, this helps programmers.

Development and Design Features

The concept was developed to accommodate a number of features for rotating applications.

* Radial adjustment is accomplished through the adjustable Kenbore heads. These field-proven Kenbore heads, in use since 1970, have large radial adjustment capabilities (Fig. 3). This means only seven sizes of bars, and a variety of heads, can cover a bore range of .500" to

4.332" diameter in the Kenbore system (Fig. 4). A bore range of .500" to 5.640" can be covered on five bars and heads in the Kenbore-Microbore system (Fig. 5). This system has a dual radial adjustment feature. It uses both the rigid Kenbore adjustment and the minute adjustment of the Microbore cartridge.

* Axial adjustment or overall length of the bar is obtained with a threaded adjustment ring (Fig. 6). The knurled ring is at the rear of the bar on the adaptor flange. Graduations on the sloped face of the ring are divided into increments of .001" to simplify axial adjustments to specified preset lengths. A #12 Acme thread on the inside diameter of the ring provides adjustment stability and strength for the axial feature.

* Chip clearance was improved on rotating Kenbore heads. This clearance is between the cutting insert radius and the side of the head. Sufficient chip clearance is essential because the distance between the bore and the head is always constant for a rotating application (Fig. 7).

* An interchangeable boring system is achieved through the bar-adaptor connection (Fig. 8). This connection makes the Kenbore and Kenbore-Microbore systems adaptable to any machining center. The bars are bored and reamed to fit over a specified pilot diameter on the front of the adaptor. The bars are locked to the adaptor as the bar's cone point lock screw engages the incline notch in the adaptor pilot. The tightening action of the screw in the bar against the incline in the pilot forces the bar firmly against the adjusting ring on the adaptor. This system can be interchanged with either existing or new machines. Also interchangeable are the bars in the Kenbore and Kenbore-Microbore systems. This means fewer bars to buy and stock.

* Point orientation is the relationship between the cutting insert point and the adaptor's drive key. This relationship permits the user to stop the machine at the bottom of a

bore, then move the workpiece away from the insert cutting point and retract the bar without scoring or marking the bore surface. Orientation is maintained by a dowel pin in the bar that fits into a keyway in the adaptor pilot. This keeps the insert in line with the drive key (Fig. 11).

Strength Evaluation

The axial connection in a rotating boring bar had to be evaluated at three highly concentrated stress areas. These areas are the bar, pilot and axial thread undercut diameters, where an approximate 4:1 ratio was maintained in relation to the bar overhang. This is usually referred to as the length to diameter ratio (L/D ratio) (Fig. 9). A 4:1 L/D ratio for a steel bar assures maximum strength for minimum deflection and stress. Deflection through the connection was further evaluated in regard to the diametral clearance and length of pilot. Deflection can be calculated from the formula of

$$Y = \frac{PL3}{3El} \text{ , or}$$

$$Y = \frac{\text{cutting pound load x overhang length}^3}{3 \text{ x modulus of elasticity x moment of inertia}}$$

and the formula for stress would be:

$$S = \frac{MC}{I} \text{ , or}$$

$$S = \frac{\text{cutting pound load x overhang length x 1/2 bar diameter}}{\text{moment of inertia}}$$

Torsional stresses on a bar should not exceed approximately 70% of the material's yield strength. The point orientation dowel pin and the encased lock screw in the pilot resist torsional movement (Fig. 11).

A static deflection comparison test was evaluated between a solid bar and a bar with the axial adjustment. Figure 10 illustrates the results of the test and indicates minimal loss of static stiffness due to the additional connection. The bar's dynamic stiffness ($Kd = 2 \alpha k$), or ability to suppress cutting chatter, is directly proportional to the static stiffness ($K = \frac{3EI}{L^3}$). A 400 pound cutting load would produce only a .0002" difference in deflection between a solid bar and a non-adjusted axial bar. A 400 pound cutting load would result in only .0005" difference in deflection between a solid bar and an adjusted axial bar. A 400 pound load would be used with a roughing cut where tolerances are not critical. A finishing cut would normally be less than a 100 pound load with zero difference in deflection between the solid and axial bars.

Locking force vectors were analyzed to provide maximum holding and resisting moments. Figure 11 reflects the plotted resisting force moments and their ability to resist unloading from various cutting loads. The forces are practically divided equally due to the driving incline at a 45° angle.

Economics

This new interchangeable tooling concept is designed to reduce tooling inventory costs drastically. Now it is possible to standardize on a series of adjustable heads and bars that are adaptable to any machining center. Savings can be quite substantial as additional machining centers are purchased, operating time can be increased, and the production floor has greater flexibility to increase productivity. This unique boring system has been well received since its introduction to industry at the 1976 International Machine Tool Show.

Acknowledgments

The author wishes to express sincere appreciation and acknowledgment to W. L. Kennicott, vice president of engineering; J. W. Heaton, engineering manager; E. L. Sorice, product sales manager; and to Kennametal Inc. for the opportunity to author and present this technical paper. Also to J. Dicesere, A. Marcinko, A. Shuster, T. Caldwell and L. McDowell, and to Kennametal drafting and advertising personnel for their help in preparing models, conducting tests, producing illustrations and typing the manuscript.

FIGURE 1

Standard
V-Flange Adapter

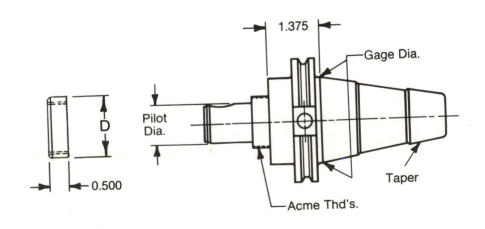

FIGURE 2

Radial Adjustment

FIGURE 3

Kennametal ANSI 45 tooling system

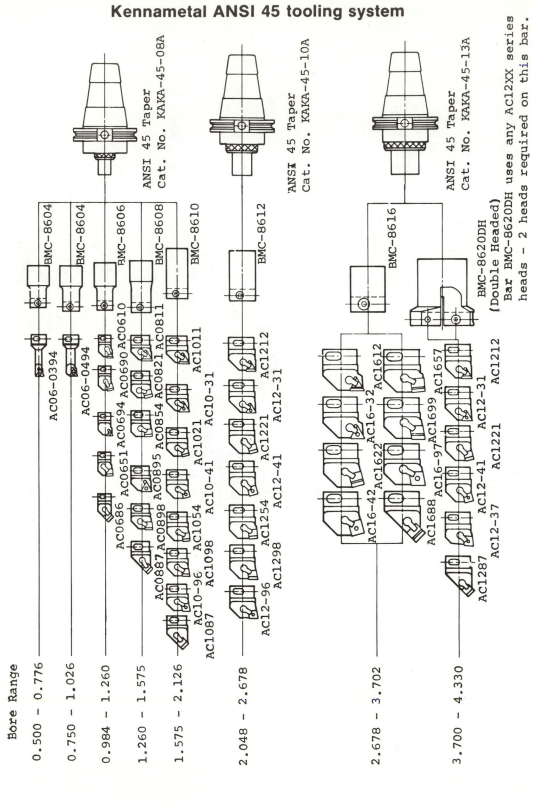

FIGURE 4

Kennametal ANSI 45 tooling system

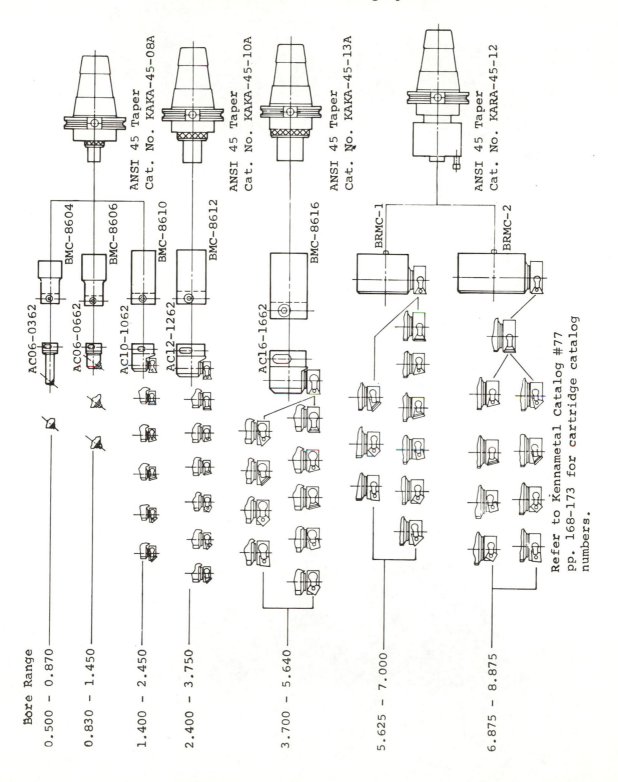

FIGURE 5

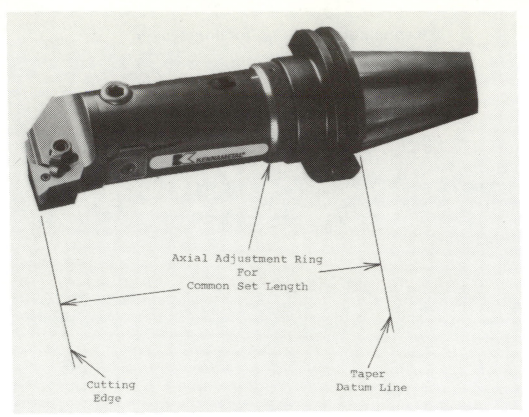

Axial Adjustment Ring
For
Common Set Length

Cutting
Edge

Taper
Datum Line

FIGURE 6

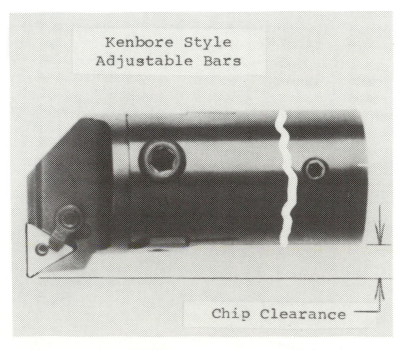

Kenbore Style
Adjustable Bars

Chip Clearance

FIGURE 7

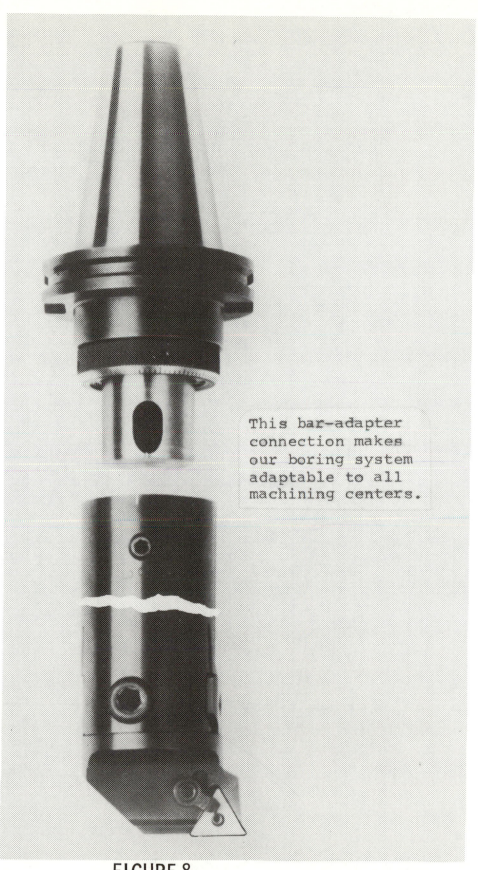

This bar-adapter
connection makes
our boring system
adaptable to all
machining centers.

FIGURE 8

MACHINING CENTER PRODUCT LINE – L/D RATIOS

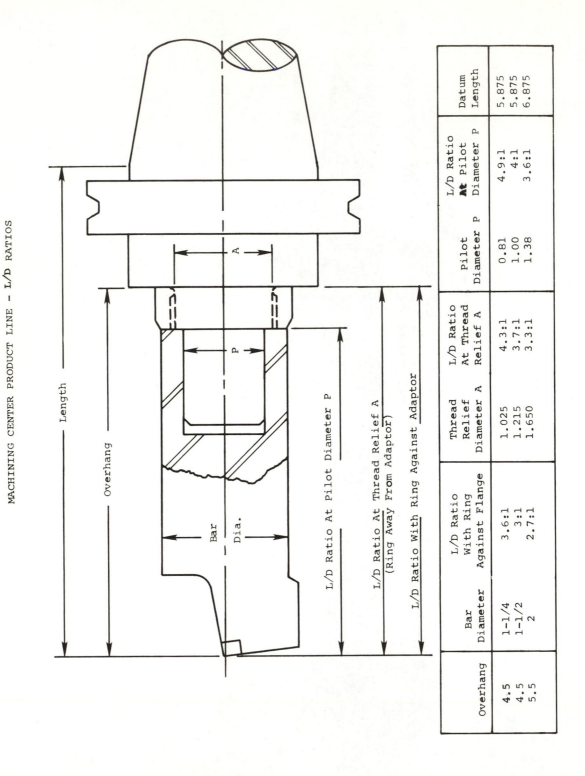

Overhang	Bar Diameter	L/D Ratio With Ring Against Flange	Thread Relief Diameter A	L/D Ratio At Thread Relief A	Pilot Diameter P	L/D Ratio At Pilot Diameter P	Datum Length
4.5	1-1/4	3.6:1	1.025	4.3:1	0.81	4.9:1	5.875
4.5	1-1/2	3:1	1.215	3.7:1	1.00	4:1	5.875
5.5	2	2.7:1	1.650	3.3:1	1.38	3.6:1	6.875

FIGURE 9

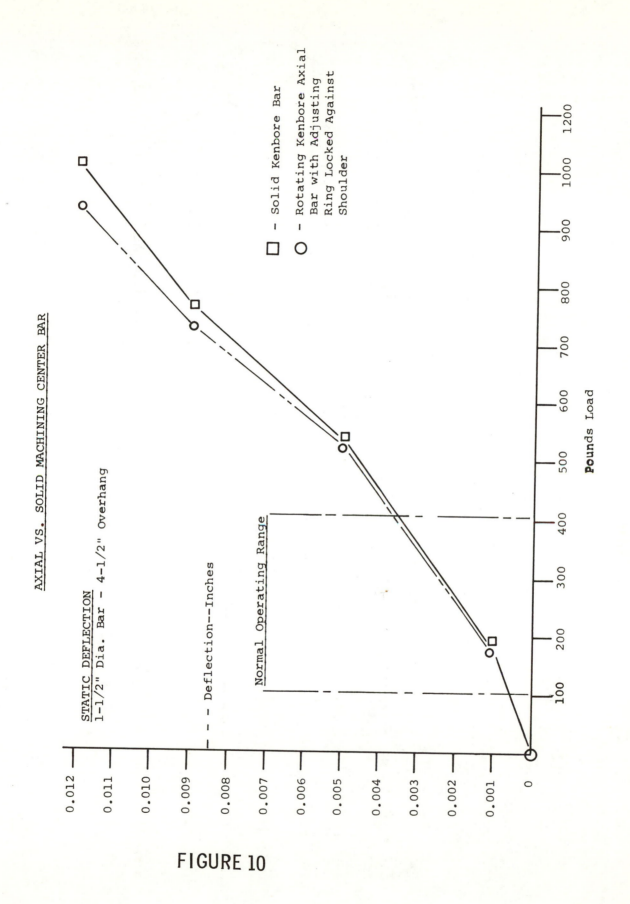

AXIAL VS. SOLID MACHINING CENTER BAR

STATIC DEFLECTION
1-1/2" Dia. Bar - 4-1/2" Overhang

− − − Deflection--Inches

Normal Operating Range

Pounds Load

□ − Solid Kenbore Bar

○ − Rotating Kenbore Axial
Bar with Adjusting
Ring Locked Against
Shoulder

FIGURE 10

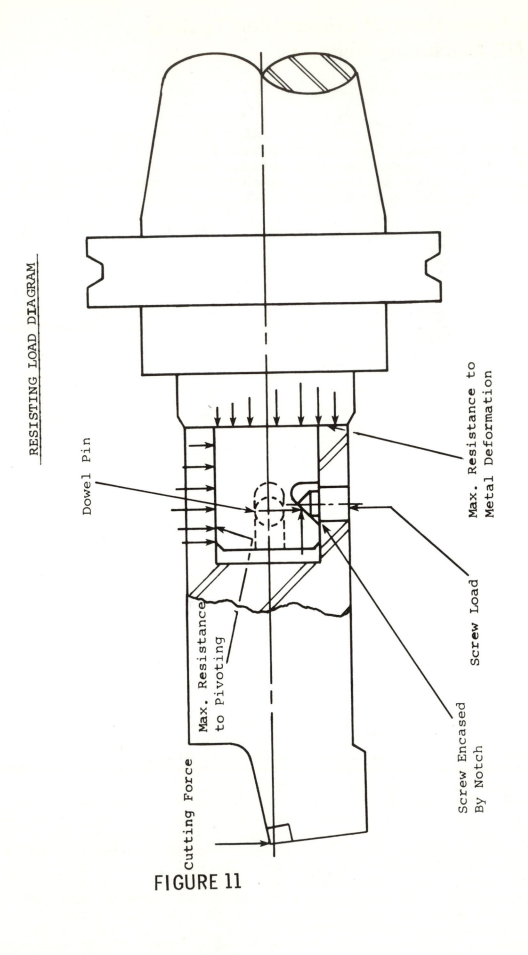

RESISTING LOAD DIAGRAM

Dowel Pin

Max. Resistance
to Metal Deformation

Max. Resistance
to Pivoting

Cutting Force

Screw Encased
By Notch

Screw Load

FIGURE 11

The Application Of Coolant Fed Tooling To NC Machining Centers

By Alan D. Glasscock
Supervisor, Customer NC Programming and Tooling
Sundstrand Machine Tool

Over the years numerous technical articles have extolled the benefits of applying coolant directly to the tool/chip interface. In the past, the application of this principal to rotating cutting tools was restricted to special tooling requiring specialized machine tools and large part quantities.

This principal has been applied to general purpose, multifunction machine tools as an extreme measure for solving the most difficult machining problems. The application of this tooling to everyday production has been avoided. Such applications require the modification of the machine tool to include a high pressure coolant pump with its associated plumbing and controls, special coolant gland tool adapters, coolant guarding, and a coolant return and settling system.

The increasing types of general purpose coolant fed tools and adapters encouraged the users and the builders of general purpose machine tools such as N/C machining centers with automatic tool changers to consider a better method of applying this tooling to their equipment.

HISTORY

In early 1963, a 5-axis Sundstrand machining center was field retrofitted with a large high pressure coolant pump. The original application was for coolant fed twist drilling deep holes in oil well instrumentation components. Since the installation of the first machine, two additional machining centers have been added, a 3-axis machining center with index table and a 5-axis machining center, all with the "coolant through the spindle" option. The application now varies to twist drilling, reaming, gun drilling, end milling, core drilling, and boring operations in steel, titanium, stainless steel, and aluminum parts. The parts are machined in a random mix of material types which limits the type of coolant that can be used.

In 1963, a manufacturer of aircraft gear boxes tested and developed tooling and processing techniques for machining magnesium gear housings on a 5-axis machining center. Included in the testing was a modified gun reaming technique called "boreaming". Boreaming required high pressure cutting oil passing through the tool center in order to remove the chips and to attain the required surface finish (16 RMS to 32 RMS).

Following the test and development period, an eight machine system was installed to machine a large variety of the magnesium housings. The system was composed of 4-axis machining centers with a special 5th axis vertical indexer and interconnected with a conveyor-type material handling system. The machines were connected to a central coolant sump, filtration, and supply system. Each machine had a hood, which was equipped

with fire extinguishers to prevent a magnesium fire. The hood
was lowered over the work zone during the machining cycle. The
machines used the recently developed "boream" tooling to machine the magnesium gear housings.

In 1974, a manufacturer of truck automatic transmissions purchased two 4-axis machining centers to manufacture aluminum
valving components. Both machines were equipped with the
high pressure coolant through the spindle and the coolant
guard options. Additionally, both machines were tooled and
programmed to use "boream" tooling.

DESIGN ACTIVITIES

As the result of the experience with these installations, as
well as several others, a design revision was made in the current small and medium sized machining center product lines to
incorporate a better coolant control system.

The original coolant guard system was designed to contain the
normal coolant spray system which pumped up to 5 gallons/
minute with a pressure of 8 pounds/square inch. The guards
were designed to contain the coolant within the coolant and
chip troughs in the machine base; thus the restricted guards
had to allow clearance gaps between themselves and the moving
axis members. No additional provisions were included to contain the high pressure coolant flowing through the spindle and
the tool at a rate of more than 10 gallons/minute at pressures
up to 800 pounds/square inch. The guards could not contain
coolant flowing at such high rates and pressures.

A revision of the design of the base of the machine was under-
taken to extend the coolant splash pan thirty inches on either
side of the machine (see Figure 1). This made it possible to
catch the coolant dripping off of the various machine components
and the overspray from the spray coolant nozzles. Additionally,
all the bases were equipped with the necessary mounting points
for the optional coolant guards; thus the guards could be field
retrofitted as desired.

The newly designed guards completely enclose the machining
area. All axis slides and their corresponding protective
covers were inside the sheet metal enclosure. The guards were
designed to fit the machine in either the pallet changer or
non-pallet changer configuration by the substitution of several panels (see Figure 2).

To date, all of the machining centers equipped with coolant
through the spindle have utilized the continuous high pressure
principal. A conventional rotary pump, pressure tank, associated hardware, the necessary plumbing and connectors, and
filtration system comprise the unit which is situated at the
rear of the machine.

The filtering system is important to the function of the pumping unit because it prevents the contamination of the tool retention and collet mechanisms in the spindle with particulate matter. Filtration will also prevent the premature failure of the coolant pump and the plugging of the coolant holes in small diameter coolant fed tooling. The filtering system is configured to permit the changing of filters without having to shut down the machine by directing the flow from one filter to the other with a selector valve while the first filter is replaced.

At the present time, a pulsating pump system is undergoing evaluation testing. The pump is capable of continuous flow, alternating mist-pulse-mist flow, and continuous mist flow. The feature of greatest interest is the mist-pulse-mist flow which aids in breaking up the vapor barrier at the tool/chip interface, and facilitates the movement of chips up the flutes of drills, reamers, and other deep hole tools [4]. The continual flow of mist can also be provided by the pump which uses shop air supply as its power source. The various coolant functions are actuated from program command, while the pressure is adjusted by a hand valve on the pump unit.

In order to facilitate easy removal of chips (steel parts in particular) from deep drilled holes, heavier than usual feeds are used which cause the chips to break up due to their increased thickness. By breaking up the chips in short, small segments the coolant flow up the flutes can carry the chips out of the hole [8]. The increased feed rates result in greater thrust forces necessary to push the chisel point of the drill into the material. The servomotor thrust capacity was increased to make it possible to take full advantage of coolant feeding drills.

The basic design of the coolant flow system through the spindle has not changed since 1962, but an explanation of how the flow of coolant from the pump through the spindle to the tool holder is necessary to understand the reason for various design decisions.

The coolant in the machine sump is pumped through a 5 micron filtering system by the spray coolant pump to the forty-five gallon tank in the high pressure pumping unit. From the pump, the coolant enters the spindle through a rotary coupling at the rear of the spindle (see Figure 3) and flows through a tubular rod through the spindle to a flanged stop button for straight shanked tool holders. The stop button is the precision gauging stop for the preset tool holders. The face of the stop button is grooved to hold an o-ring, which seals the coolant flow from the spindle into the tool holder.

The tool holder has an adjustable stop screw with a flange and a hole through the center of it (see Figure 4). The flange seats against the o-ring to prevent coolant leakage and pressure loss while the coolant passes through the stop screw into the holder through a check valve and into the tool.

For No. 50 taper "V" flange tool holders (see Figure 5), the coolant passes through a tubular knockout rod, the axial retention collet, the tool holder retention stud, and into the holder and the tool.

TOOL HOLDER CONSIDERATIONS

Tool holders that accept straight shank tools are better suited for coolant fed tooling. Solid type holders, such as end mill and automotive shank, do not require special consideration for sealing. The collet type holder, on the other hand, must be sealed in some form to prevent excessive leakage and the corresponding pressure drop (see Figure 6). Two methods of sealing collet holders that should be considered are sealing the slots in the collets with an epoxy or sealing the end of the tool against a Teflon-like faced screw inside the holder. The sealing screws can be purchased from its manufacturer or from some tooling suppliers. Collet slots can be sealed by various collet manufacturers.

FIXTURE DESIGN CONSIDERATIONS

The design of fixtures for N/C machining centers using coolant fed tooling requires the consideration of allowing for unrestricted return coolant flow to the machine sump. Fixtures should have cutouts of adequate size to allow the free flow of coolant and chips without the chips damming up the opening. Sheet metal baffles and covers can also be used to facilitate the free flow of chips and coolant to the machine sump.

Frequently, the type of tooling employing the coolant feeding principle are drills with high length to diameter ratios. The design of a fixture for a part requiring a long drill could include a provision for a quickly mounted bushing plate which can be installed during the machining cycle (see Figure 7).

PROGRAMMING CONSIDERATIONS

If the location of the hole is not critical, the bushing plate may be eliminated in favor of programming a stub screw machine drill to drill a shallow lead hole for the long drill. The stub screw machine drill should have the same or a flatter point angle than the coolant feeding drill and should drill the hole to a depth of two times the diameter of the drill [8].

Drilling of a lead hole may not always be necessary, but spotting of the hole should be employed to insure an accurate start. Again, the point angle of the spotting drill should match or exceed the point angle of the coolant feeding drill [9].

By grinding the same or greater point angle on the lead or spot drill, the concentration of the stresses at any one point on the lip of the coolant feeding drill is eliminated. This practice is especially important for drilling high strength and exotic metals [8].

When drilling a through hole, the feed and speed for the drill should be reduced just before it breaks through because the drill will again lose the benefit of cooling the lips of the drill and the flushing of chips [9].

CUTTING FLUID SELECTION

Selection of cutting fluid type is usually based on the effects that the specific fluid has on machining when it is sprayed on the cut. In the case of high pressure coolant applications, several other factors must be considered. Cutting fluids can usually be categorized as mineral oils or water soluble coolants. Each of the fluid types has generalized characteristics which must be considered when making the cutting fluid selection.

Cutting oils offer good to excellent lubricity. In severe machining operations, the lubricity and the ability of oil to maintain an oil film under extremely high pressure is desirable.

The effects of forcing the cutting oil through a small opening, an orifice, causes the oil to atomize. The misting of the oil can make possible the contamination of the air in the area around the machine. Special venting procedures may be necessary to maintain safety requirements. Additionally, the misted oil is finely mixed with the air and this combination is a potential fire hazard due to the mixture's low flash point. Certain fire protection procedures may have to be instituted [7].

Another characteristic of cutting oils which should be considered is their comparatively low thermal conductivity. Oil is slow to take the heat out of the tool/chip interface and slow to give up the heat. The tendency to hold heat is not conducive to good tool life where heat reduction is needed [4 and 8]. This condition could be significant enough to require the coolant pump system to contain an oil chiller to maintain an acceptable oil temperature [7].

Water soluble coolants, by their nature, are much less of a fire hazard by virtue of the fact that they are somewhere between 80% and 95% water. Even in an atomized state, the likelihood of a fire is considerably reduced.

Water has a relatively high thermal conductivity rate which makes water soluble coolant an attractive selection for machining conditions that require the conduction of the heat

released by chip formation away from the tool tip [4].
(Machining shallow holes in exotic metals for example [8].)

Because of the low percentage of concentrate in water soluble
coolants, the lubricity is significantly less than the lubricity of cutting oils. Pumps used for coolant through the
spindle require a minimum of 5% of the coolant be soluble
concentrate. Any percentage less of concentrate could do
serious damage to the pump.

FEEDS AND SPEEDS GUIDELINES

Sources of feeds and speeds guidelines can be found in the
Machinability Data Center Handbook Ed 2 covering oil hole
drills and reamers, the George Whalley Co. Coolant Fed Tooling
Catalog covering coolant feeding drills, the sales brochure
for Wheel Abrator-Frye's Division's Jet Pulser systems covering coolant feeding drills, and the Eldorado Gun Drilling sales
brochure covering gun drilling. Most manufacturers of coolant
fed tooling will be happy to discuss specific applications of
their tooling to production parts.

POTENTIAL APPLICATIONS

Applications to consider for coolant fed tooling are subject
to numerous evaluation criteria, but listed below are a number of likely candidates:

1. Deep holes that do not require good finish (125 RMS) are
    good candidates for coolant feeding twist drills and
    straight flute drills with either high speed steel or
    carbide tips.

2. Shallow holes (1 to 6 diameters deep) where either tool
    life, cycle time, or chip removal is unsatisfactory, are
    good candidates for coolant feeding twist drills and
    straight flute drills with either high speed steel or
    carbide tips.

3. Tapped holes, where chip removal or tool life is unsatis-
    factory, are good candidates for coolant feeding taps.
    NOTE: Coolant feeding taps and tap drills can eliminate
    the need to interrupt the machining cycle to clear chips
    from a hole before it is tapped. This is of particular
    importance where one operator runs several machines or
    on a computer controlled manufacturing system where one
    or two operators run several machines and a material
    handling system.

4. Pockets or slots, where chip removal or tool life is
    unsatisfactory, are good candidates for coolant feeding
    t-slot cutters and end mills.

5. Holes under 1.625" in diameter that require good finishes (better than 125 RMS) and size control, are good candidates for gun drilling, "boreaming", or reaming. NOTE: The selection of the specific method will be influenced by hole depth, part material, and part configuration.

6. Holes from 2.0" to 3.5" in diameter that must be made from solid and are more than 1 diameter deep, are good candidates for spade drilling. NOTE: The thrust capacity of the machining center may limit the maximum size of spade drill that can be used. The thrust of a spade drill can be greatly reduced by drilling a small diameter lead hole for the spade drill. Trepanning may also be used for this type of operation. NOTE: Consult a trepanning tool manufacturer before implementing trepanning.

7. Back spotfacing holes where surface finish, tool life, or chip removal is unsatisfactory, is a good candidate for automatic back spotfacers. NOTE: One automatic back spotfacer on the market now is actuated by high pressure coolant passing through the tool's center.

8. Recessing grooves in bores where surface finish or chip removal is unsatisfactory, or tool breakage is encountered is a good candidate for coolant feeding recessing tools.

CONCLUSION

The ability to perform the previously mentioned operations on a machining center that is designed to accommodate coolant fed tooling, gives the capability to use tooling and methods that were not previously available to the user of general purpose, multifunction machine tools. By applying these methods to the many parts that are or could be machined on machining centers, the floor-to-floor time for the machining cycle will be decreased and the part quality improved.

BIBLIOGRAPHY

[1] Bloch, Frederick S. and Bloch, Eric, "Think Shallow-Hole Gun Drilling", Modern Machine Shop, November 1972, Page 80-9.

[2] Brockman, R. W., et al, "Drill Design and Application Requirements for Optimum Coolant Feeding Twist Drill Usage", ASTME Technical Paper MR67-104 (1967).

[3] Donovan, Ray B. and Rich, E. A., "Process Report--The Revolution in Drilling Technology", ASTME Technical Paper MR68-504 (1968).

[4] Gettleman, Ken, "Drilling Productivity Can Be Improved", Modern Machine Shop, March 1969, Page 130-142.

[5]   Rich, E. A., "Effects of Pulsating Coolant Pressures in Oil Hole Drills", ASTME Technical Paper MR66-186.

[6]   Rich, E. A., "Progress Report on Pulsating Coolant Pressures in Oil Hole or Coolant Fed Drills", ASTME Technical Paper MR67-108 (1967).

[7]   Rich, E. A., "The Growing Acceptance of Coolant Fed Drilling and Cutting Tool Systems for Drastic Cost Reduction in Metal Working", ASTME Technical Paper MR69-173 (1969).

[8]   Trainor, Ray, "Drilling Exotic Metals with Coolant Fed Drills".

[9]   Trost, Charles J., "Keys to Better Deep-Hole Drilling", Modern Machine Shop, February 1975, Page 94-8.

[10] "Tough Drilling Problems? Gun Drilling Can Bail You Out", Metalworking, November 1968, Page 35-7.

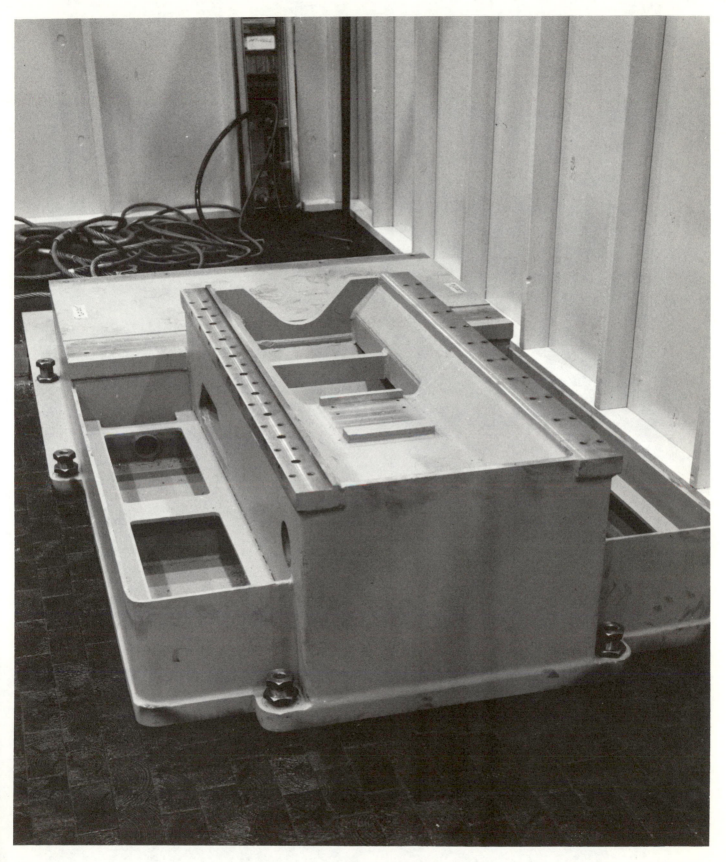

Figure 1A: "Old Style Base"

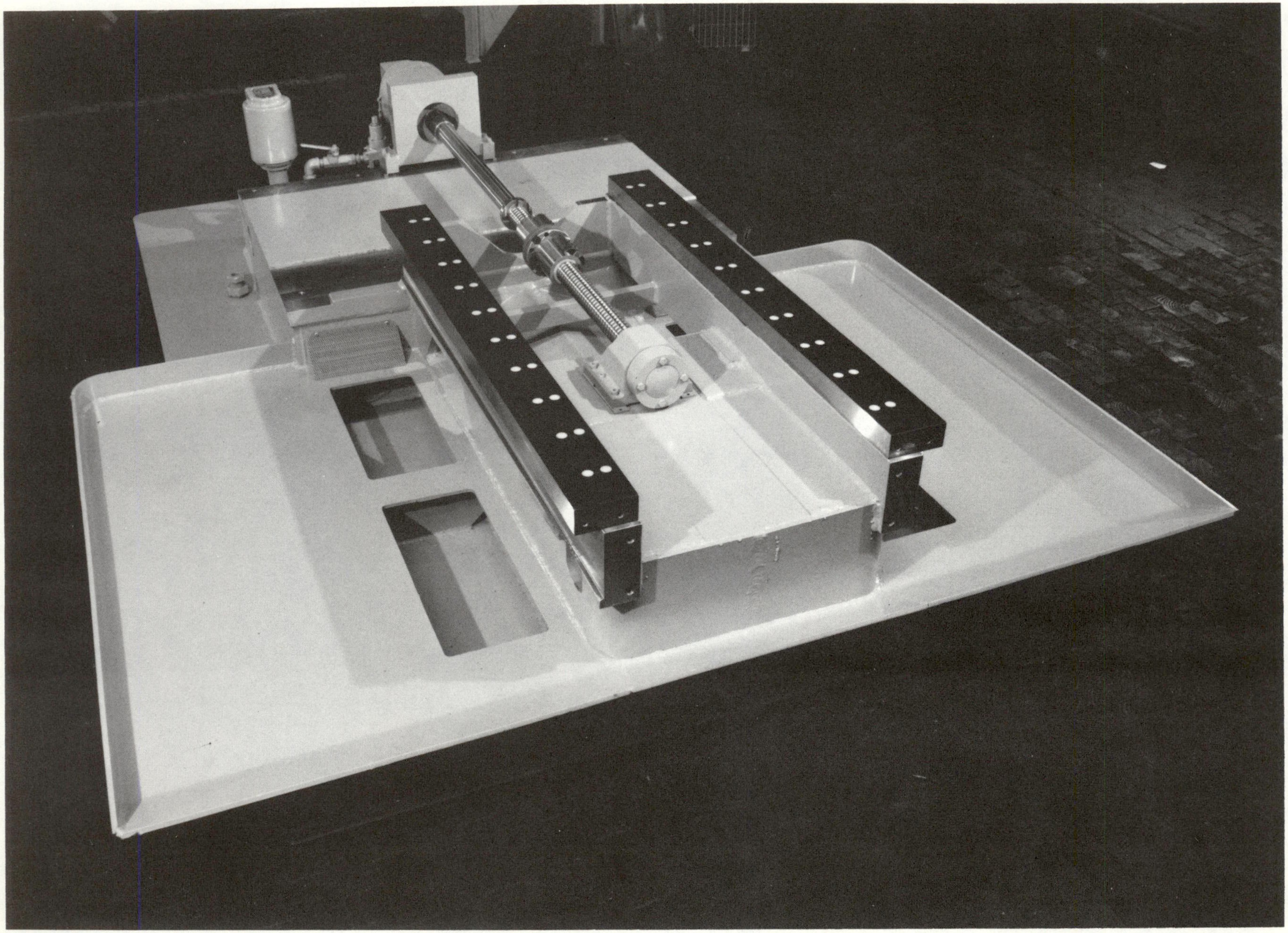

Figure 1B: "New Style Base"

Figure 2: "Non Pallet Changer Guards"

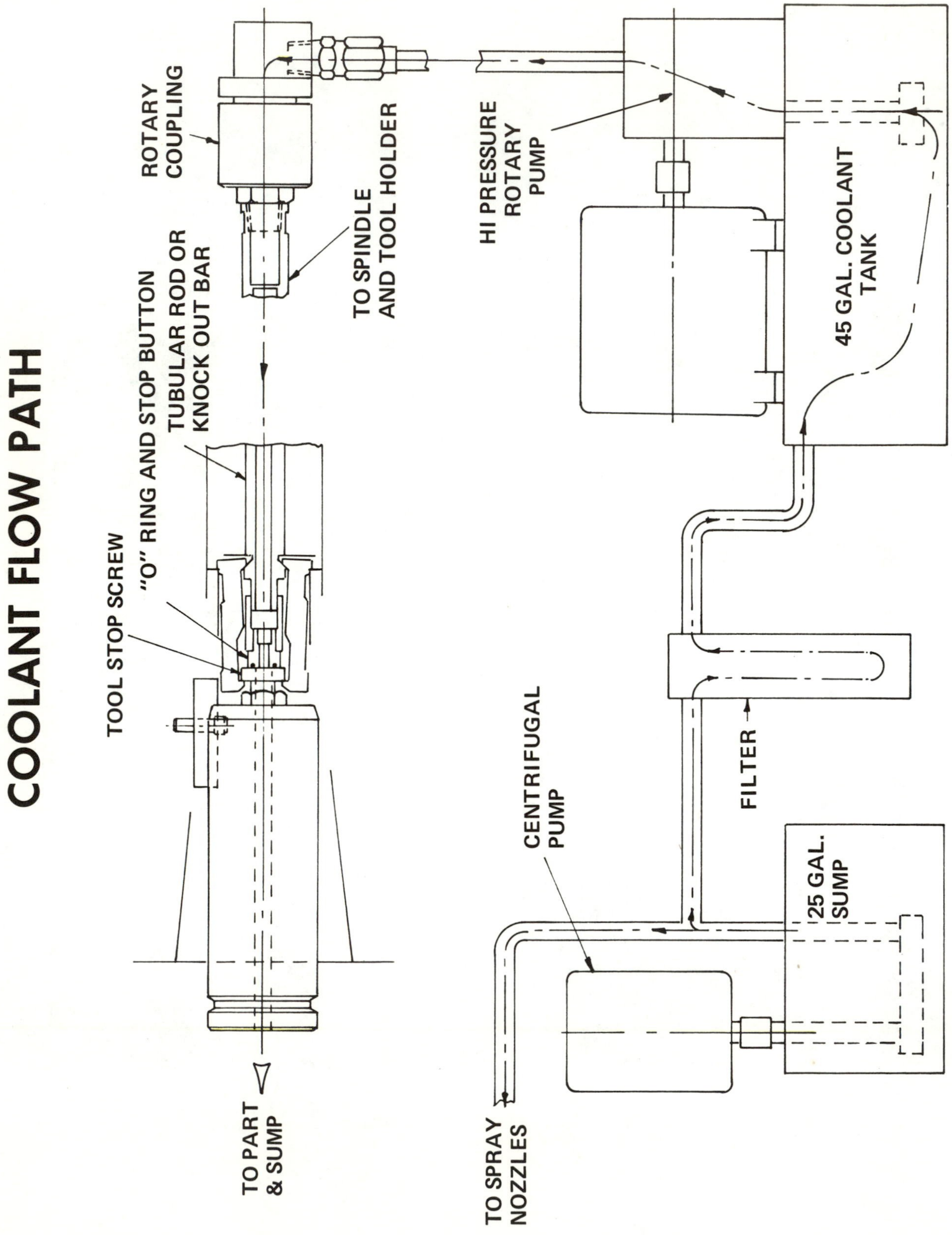

COOLANT FLOW PATH

ROTARY COUPLING

TO SPINDLE AND TOOL HOLDER

HI PRESSURE ROTARY PUMP

45 GAL. COOLANT TANK

"O" RING AND STOP BUTTON

TUBULAR ROD OR KNOCK OUT BAR

TOOL STOP SCREW

CENTRIFUGAL PUMP

FILTER

25 GAL. SUMP

TO PART & SUMP

TO SPRAY NOZZLES

Figure 3: "Coolant Flow Path"

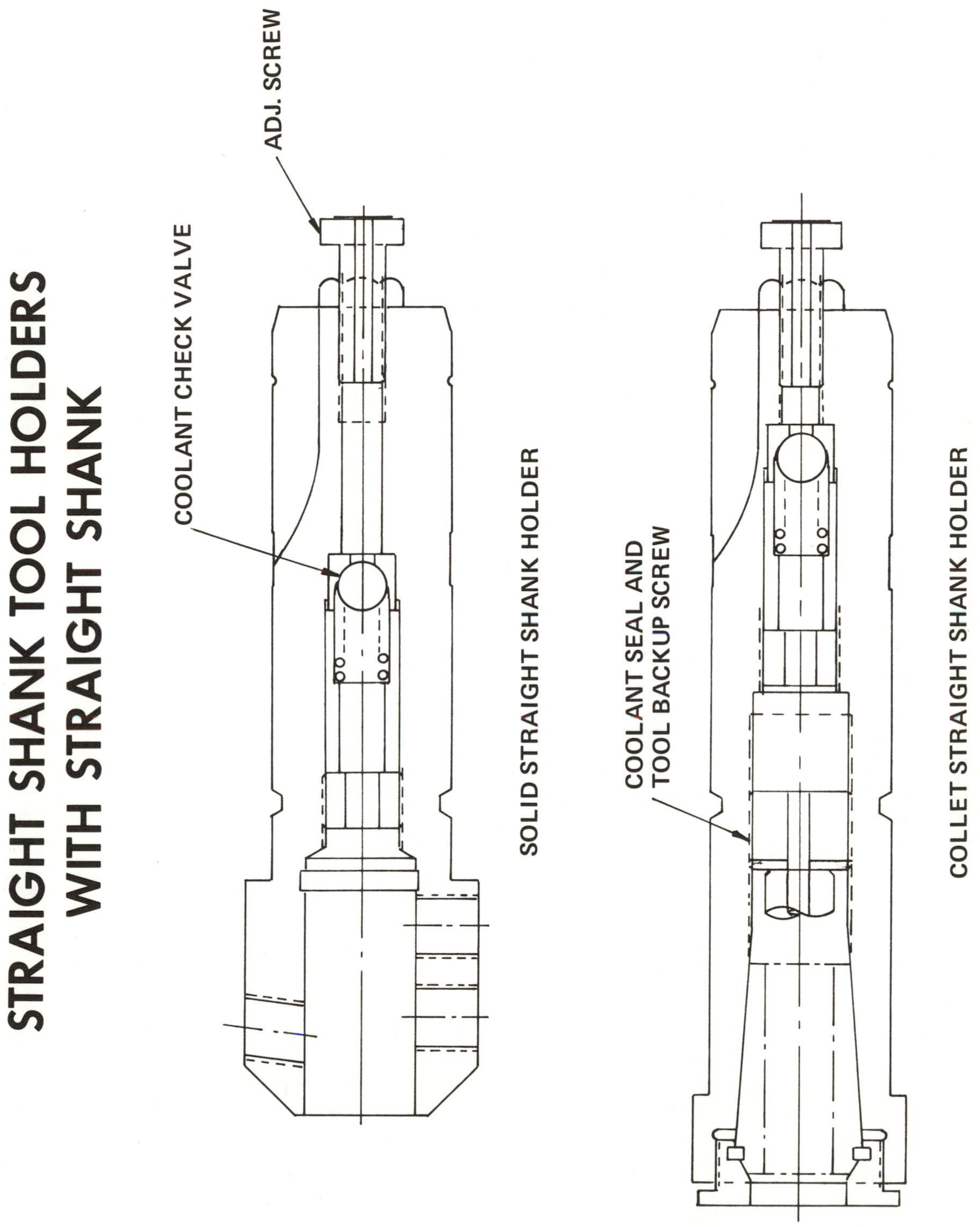

STRAIGHT SHANK TOOL HOLDERS WITH STRAIGHT SHANK

ADJ. SCREW

COOLANT CHECK VALVE

SOLID STRAIGHT SHANK HOLDER

COOLANT SEAL AND
TOOL BACKUP SCREW

COLLET STRAIGHT SHANK HOLDER

Figure 4: "Straight Shank Tool Holders with Straight Shank"

163

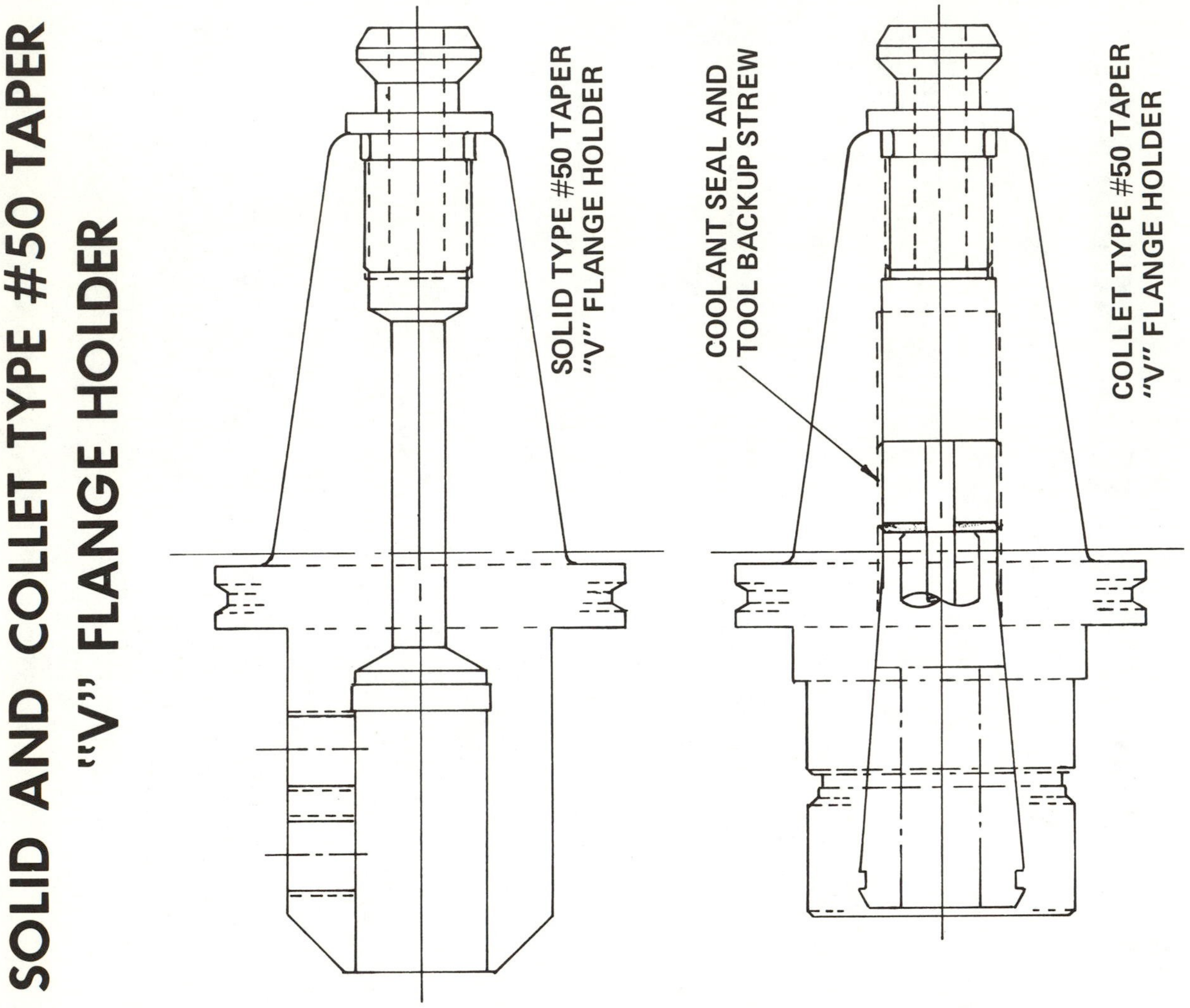

SOLID AND COLLET TYPE #50 TAPER "V" FLANGE HOLDER

SOLID TYPE #50 TAPER "V" FLANGE HOLDER

COOLANT SEAL AND TOOL BACKUP STREW

COLLET TYPE #50 TAPER "V" FLANGE HOLDER

Figure 5: "Solid and Collet Type #50 Taper 'V' Flange Holder"

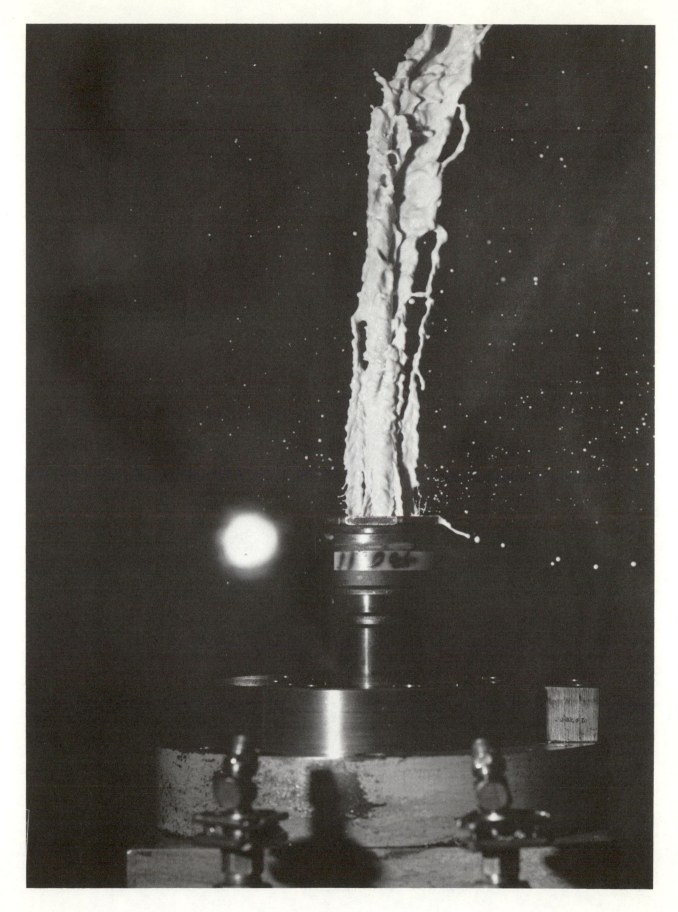

Figure 6A: "7/16" Dia. Coolant Fed Drill in Unseated Coolant Holder"

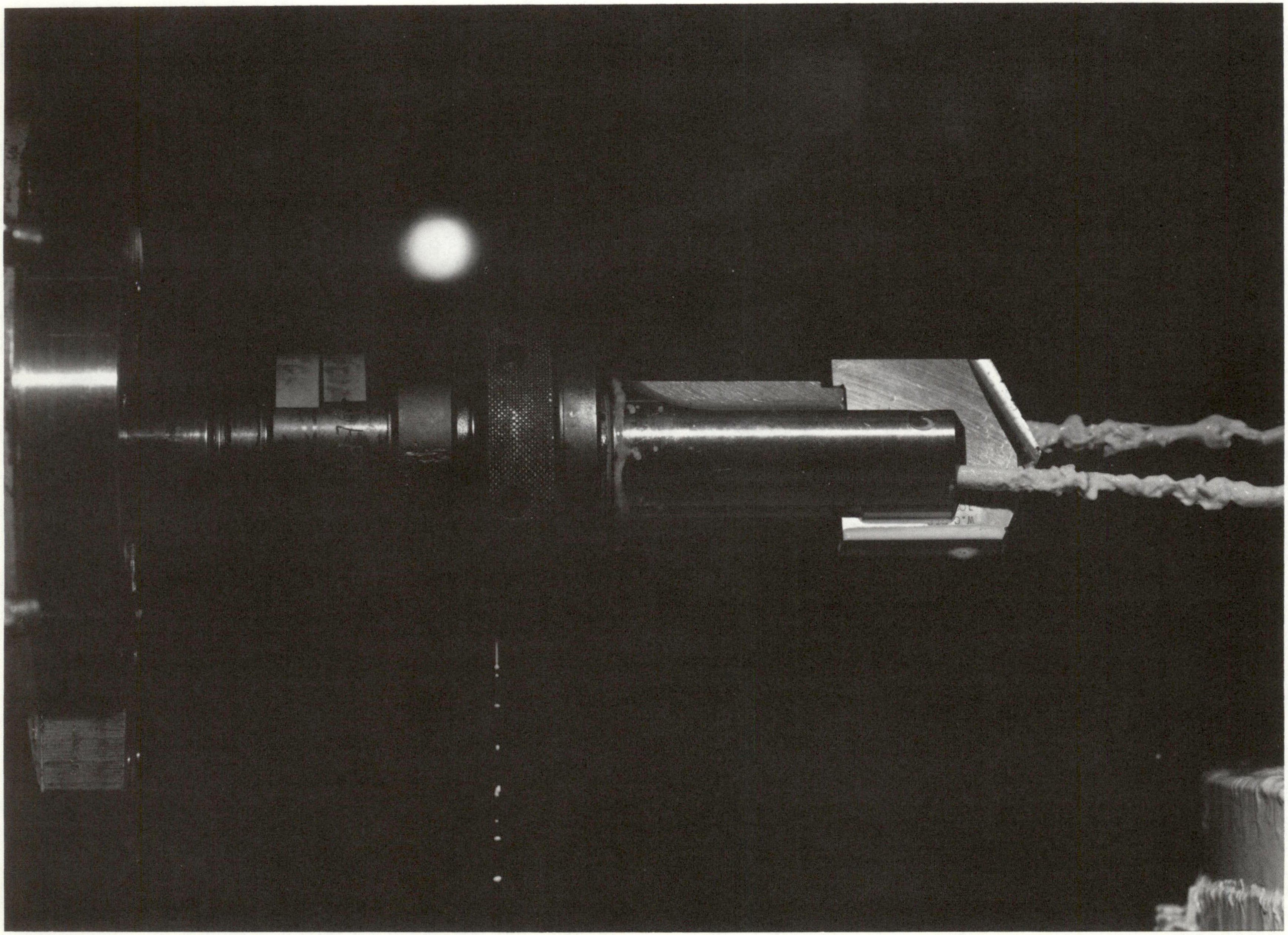

Figure 6B: "2 1/2" Diameter Coolant Fed Spade Drill in Sealed Collet Holder"

FIXTURE WITH REMOVABLE
BUSHING BLOCK

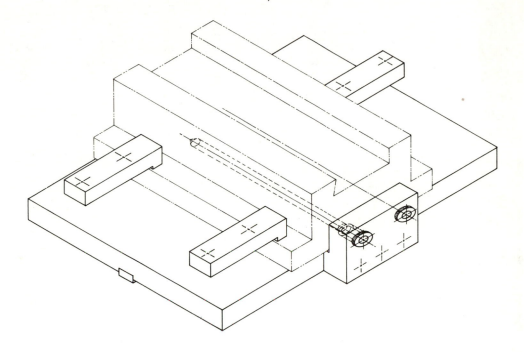

Figure 7:   "Fixture With Removable Bushing Block"

CHAPTER 5

PRODUCTIVITY

The NC Approach To Maximizing Productivity

By Henry Brace
Manager, Manufacturing Engineering
Documation Incorporated

Numerical Control when maximizing productivity cannot be thought of as simply the machine tools that are controlled by punched paper tape. It must be approached as a total system, consisting of a variety of applications in which numerical data is controlling a manufacturing process. These applications can be as simple as a digital readout system or as complex as a Computer Numerical Control machining center. This paper will address the effective utilization and integration of each of these systems in a total manufacturing system.

INTRODUCTION

This paper is based on experiences encountered in a medium size manufacturing company, producing highly sophisticated electro-mechanical high speed line printers and card readers for the computer industry.

The manufacturing cycle is controlled in the following manner. The parts necessary to assemble all of the finished products are ordered from manufacturing based on a marketing forecast. They are then put into a stockroom and the finished product is assembled based on the receipt of a purchase order. This system allows some economy in manufacturing due to quantities and responsiveness to customer needs with short delivery of finished products.

This batch process type of manufacturing differs from a job shop machining environment or a highly automated manufacturing company producing high volumes of a particular product. This is significant since many of the approaches to manufacturing problems would be entirely different in those three manufacturing situations.

A common job shop approach to many manufacturing problems must include a very high degree of versatility, they never know what their next job will be. They must therefore utilize systems or machines that will cover all the bases, or do any job that comes in the back door. Many of these shops specalize in certain types of machining or manufacturing applications but they still are not sure what their next job will involve.

The high volume manufacturer looks at his machining problems in a different light. He knows what he will be making, generally for a long period of time. His product is generally stable and mature (not subject to excessive engineering changes) before he invests the money to tool up for production. He is looking for high production specialized processes, versatility in the machinery that he buys is usually not a significant factor. He wants the ultimate production process to produce a given workpiece at minimum cost for a long period of time.

The area between these two ends of the continuum is the batch type

171

small to medium size manufacturing company. This type of company will generally manufacture its products from preproduction stages (quantities of one-ten pieces) through production stages (quantities of fifty to several hundred per month) in the same manufacturing facility. This manufacturing facility must offer the best of two worlds. It must be versatile enough to manufacture the prototypes on a timely basis, plus it must be able to offer the economies of producing in small to medium production batches.

In the early stages of a new product, the manufacturing company must be flexible enough to be able to handle a relatively high number of engineering changes with a minimum cost for retooling. As the product matures and the volumes increase they must be capable of retooling to minimize manufacturing costs.

There must be compromises made at all steps in this process.

Production control people will say they never want to be out of a particular part when it is needed on the assembly line. Financial people will say that they want to run the business with the least possible amount of money invested in inventory. Manufacturing people will say they want to make as many parts as possible everytime they setup a particular job. This amortizes their setup cost over more parts and drives down their piece part cost.

The performance of these different departments heads is not measured by the same yardstick. The production control manager is most effective when there are minimum part shortages or stoppages of the assembly line. The inventory control manager is measured by the number of times he turns his inventory every year. He is doing his best possible job when he maintains the lowest possible dollars invested in inventory. The manufacturing manager's competence is measured by his cost of producing the necessary parts and meeting the delivery dates established for those parts.

The ideal situation from an overall economic standpoint would be that you never have any parts (representing dollars invested) in inventory, but everytime the assembler needs a particular part to go into an assembly, the manufacturing guy has just finished it. No dollars invested in work in process inventory and never a shortage on the assembly line.

Since the above mentioned ideal situation is not achievable, a compromise must be reached between the production control, financial and manufacturing people that most closely fits all of their needs. This is done by establishing an EOQ, economic order quantity, for every part that is manufactured on an individual basis. Normally your most inexpensive parts are run in large quantities and stored in the stockroom, while your most expensive bulky items are run in smaller quantities and processed more frequently.

A full service total manufacturing company that wants to keep as much control over its manufacturing process as possible must, therefore be very flexible. It must be capable of producing high volume inexpensive parts such as powdered metal, injection molded plastic and sheet metal stamped parts. It must also be able to produce the expensive machined parts that are usually made on sophisticated N/C machinery.

Documation is this type of company. A vertically integrated company that produces 95 percent of their precision mechanical components in-house. Although this requires a tremendous investment in capital equipment, it gives us the ultimate in control over the product, from the original design to the pricing and delivery of finished products.

Because of the vast number of different processes and the varying lot sizes of the parts produced, Numerical Control in its broadest sense, is the most important single factor in our manufacturing operation. Numerical control and computer control is used in virtually every aspect of our manufacturing operation, from a cutoff gauge that measures the parts off our automatic saw to the computer generated manufacturing process plans.

This paper will attempt to explain how each of these systems work independently and how they are integrated into the total manufacturing system at Documation, Inc. Listed below are the topics to be covered individually:

a. Digital Readout system
b. Digital Programmable Controllers
c. Digital Programmable Back gauges
d. Numerical Control Wire cutting EDM
e. Numerical Control Turning
f. Computer Numerical Control Machining Centers
g. Numerical Control Punching
h. Computer Aided Inspection
i. Computer Aided Parts Programming
j. Computer Generated Process Plans

Figure 1

A.) DIGITAL READOUT SYSTEMS

Digital Readout Systems have proven to be a very inexpensive way to improve production and significantly improve the quality of the parts that we produce. Virtually everyone of our Bridgeport milling machines have a digital readout system installed on it. These machines are used in our production milling department, engineering model shop, and our tool and die shop. We also have these units installed on a crush form surface grinder and a gundrilling machine.

One of the big factors in the increase of productivity due to these Digital Readout Systems, is the ease of operation. The operator does not have to count the revolutions of the screw, then read the dial, to determine the distance he moved the table. The exact position of the table is constantly displayed in front of him. It instills confidence in the operator, he doesn't have to worry about whether he made the correct move. He can also use the exact print dimensions and save much time in calculating the machine coordinate dimensions to match the print coordinate dimensions.

The quality of the workpiece is significantly improved due to the fact that the digital readout systems use a non-contact linear scale. This is replacing a measurement system which is using the pitch of a lead screw which physically moves the table. The screw is subject to wear and back-lash. If it is not constantly maintained and calibrated the accuracy will suffer. The non-contact linear scale is not subject to wear and will be as accurate in five years as it was the day it was installed on the machine

Figure 2

B.) DIGITAL PROGRAMMABLE CONTROLLERS

This is our answer to the parts that are too simple for N/C or that do not run in quantities that warrant setting up a N/C machine. The pre-production run of a new product or the first few runs of the production batch is many times run on these machines instead of our fully controlled N/C machines.

When using these machines with programmable controllers, the machinist can decide how he wants to machine the part and then dial into the control the necessary X-Y machine motions to perform the operations. In effect he is a combination machinist-N/C programmer.

The controllers are simple two axis point to point systems with straight line milling and point to point positioning capabilities. The only data that the operator has to enter is the X and Y positions, the feedrate and a code that tells the machine to either stop at the end of the move or continue automatically with the next motion. The stop code would not be used if he wants to do continuous milling operations on the workpiece but would be used if he wants the machine to just position to a location so he can drill, tap or bore a hole. At the end of that operation he would simply press the cycle start button and the machine will position to the next hole.

There are programmable controllers available today that have many more capabilities than this, but in our operation the parts that require more complexity would go on our regular N/C machines.

The reason we selected this particular type of programmable controller is that it is true closed-loop system. It uses the same non-contact linear scale measuring system as our Digital readout systems. The accuracy of the system is not dependent on the accuracy of the lead screw. Because it uses the same identical scale system as our digital readouts, we can switch a machine from a Digital Readout System to a Digital Programmable Controller System and vice-versa in about fifteen minutes. We also have interchange-ability of many of the spare parts within the systems.

Figure 3

C.) DIGITAL PROGRAMMABLE BACK GAUGES

Because of the nature of our product, computer components, we have a very extensive sheet metal fabrication operation. We attempt to integrate N/C application in this area just as we do in the machine shop. The programmable back gauge systems are the most simple of the N/C applications in this area. They are installed on press brakes to automatically position the back gauge when forming sheet metal panels. We use these systems for forming all of our computer cabinets and other formed sheet metal parts.

The system is completely controlled by the machine operator. He simply determines the sequence in which he wants to make the necessary bends and the distance from the controlling surface to each bend. Once he has determined these distances, he simply dials them into the controller and everytime he cycles the press brake, the gauging system will automatically position to the proper location for the next bend.

The way the press brake would normally be setup without a programmable back gauge would be to use a micrometer type standard back gauge. Using this system the operator would determine the distance from the controlling surface to his first bend. He would then make this setting on the machine and make a test bend, check the part, make the necessary corrections and then perform that operation on all of the parts in that batch. He would then repeat that process for each bend that is necessary in the finished part.

With the Programmable Back gauge he only handles the part one time. When he picks up the piece of sheet metal, he does all of the forming before putting the part down. This has proven to be a tremendous increase in productivity over the method we used before we purchased the Programmable Back gauges. The other benefit we derived was in the quality of the finished part. The systems use a ball-screw positioning mechanism that assures extreme accuracy and repeatability. If the operator finds he has miscalculated the position of a bend, it is very easy to adjust the system to correct it, even in thousandths of an inch. Since he does all of the bends on one part before he goes to the next part he can also check all of his bend allowances and blank development. Using this method he does not make three bends on all of the parts, only to find out that the fourth bend will not work.

Figure 4

D.)   NUMERICAL CONTROL WIRE CUTTING EDM

As mentioned previously, numerical control and computer control is used in virtually every aspect of our manufacturing operation. This particular system was purchased to be used mainly in our toolmaking activities.

The process is a combination of two highly sophisticated technologies. A numerical control contour generating system and an electrical discharge machining process. Using the normal electrical discharge machining (EDM) process, the machinist must start by making an electrode which is a duplicate image of the hole or cavity that he wishes to produce. This is made from a conductive material such as copper, brass, bronze or carbon. The electrode is the tool used to produce the hole or cavity and like any tool it wears out as it is being used. The normal EDM application will always require multiple electrodes because of the wear factor involved. When the electrical charge is put through the electrode to burn or spark the material away, it also burns or sparks the electrode away.

Because of the high cost of producing the electrodes and the time cycle required, this process is generally used only when all other methods of machining have proven unsatisfactory. It is an expensive, slow operation as compared to most other machining processes.

The advent of the wire feed EDM is making this type of machining much more universally acceptable. It is no longer thought of as a last resort operation.

The Numerical Control Wire Cutting EDM has eliminated the need for the expensive consumable electrodes. Using a standard N/C contouring approach, you program the shape of the hole or cavity and a traveling spool of copper wire carries the electrical charges and burns or sparks the material away. This technique can be compared to a band sawing operation except it uses a very thin (.008) diameter wire and is extremely accurate. It will also cut any material that is a conductor of electricity regardless of the hardness. Wire feed EDM will not replace or obsolete the conventional EDM machines since it has one major shortcoming, whatever you are burning must be burned all the way through the part. It cannot burn blind cavities or cavities that only go part of the way through a block of steel. The Wire feed process has effectively eliminated the need for an electrode and eliminated the problem of size control due to the electrode burning away, but it is still a relatively slow process (about 1 inch per minute in 1 inch thick steel). The major emphasis by the manufacturers at the present time is to modify the power supply and improve the machines to increase the cutting speed.

This process, N/C wire feed EDM, has had a very significant impact on our company, far beyond the tooling programs for which it was originally purchased. One example is on intricate sheet metal parts for which we usually build stamping dies. When we first get the design of the part we will make an N/C tape to burn the first batch of prototypes on the wire feed machine. These parts are then completely finished and given to engineering for evaluation. At this point they still have the opportunity to change the part if necessary without tremendous tooling expense. If an engineering change is necessary, we can simply change the tape to make the new part. If the part is OK as designed, we simply use the same tape to burn the punch and die steels for the stamping die. At that point the time to make a die is greatly decreased. The process has eliminated all of the expensive form grinding and fitting of the die steels.

Figure 5

E.) NUMERICAL CONTROL TURNING

At Documation we have one numerical control lathe. The majority of our production turning is of the automatic screw machine variety, therefore virtually none of the jobs could be made economically on an N/C lathe.

This machine, like the wire feed EDM, was purchased for a tooling application to make crush rolls for a newly acquired crush grinder. We needed about 100 crush rolls at a purchased cost of $1500 each, rather than spend the money to buy the crush rolls, we bought the N/C lathe (a $50,000 investment) and in three months we had finished all of the crush rolls. We now use the machine to re-cut the rolls as necessary and we are redesigning some of the parts in our product to make them suitable to manufacture on the N/C lathe. This is a case where we can redesign our product to cost reduce it and utilize a manufacturing process that is available.

Figure 6

F.)   CNC MACHINING CENTERS

Our machining centers are by far our largest investment in N/C equipment. We are presently operating 18 N/C and CNC machining centers making precision machined parts for our high speed line printers and card readers.

In the purchase of N/C machinery we have a number of strict requirements. One is that tooling must be compatable with similar machines in the shop and that the machines, if possible, have a true closed-loop feedback system, such as Inductosyn linear scales on all three axes.

Because of our strict requirements for common toolholders, we are able to run 16 of our machines on just 2 types of toolholders, #40 and #45 taper shank holders. The only thing that varies from one machine to another is the retention studs and we can change those in the holders. This has limited our total investment in toolholders to less than $40,000. It also allows us much flexibility in changing jobs from one machine to another, the cutting tools are not limited to one machine.

At the present advanced stage of N/C technology, the biggest productivity gains are going to be made in spindle utilization. For our type of work we are not going to find significant productivity gains through improved speeds and feeds of cutting tools, nor through faster positioning speeds of the machine. The most significant gains must therefore come from the material handling aspects and the fixturing of the parts. The key to increased productivity is multiple station fixturing, getting several parts in front of the spindle at one time and shuttle type fixtures where you can load and unload parts while the machine is running. We recently realized a 100 percent productivity improvement on one of our parts by simply changing the job from one machine to another. The speeds and feeds are no different and the new machine is no faster than the old one. The only difference is the fact that the new machine has a pallet shuttle system. This allows the operator to unload, clean and reload the work holding fixture while the machine is running a different part. What this proved to us was that we were spending as much time loading and unloading as we were machining.

Another area that has contributed to increased productivity is in cataloging and numbering all of our tools. All tools are preset and delivered to the machines with the necessary fixtures and tapes to help minimize down time.

Figure 7

G.) NUMERICAL CONTROL PUNCHING

At Documation we feel that our Numerical Control Punching machines are the best 'short-run stamping system' available. We have four different approaches that we take to our manufacturing of sheet metal parts. Which of the approaches we use is determined by the size, quantity, complexity and our confidence that the part will not change due to an engineering change order.

The first of these four approaches is the straight N/C approach using all standard tooling and running multiple pieces on one sheet of material. While we are punching the parts we also punch tooling holes, at a known location, to be used for subsequent shearing operations. The shearing operation uses a pin-gage shear, which has two micrometer adjustable pins to locate the punched blank relative to the shear blade. This has proven to be extremely accurate since you don't accumulate error due to back gauging off a sheared edge. This approach is taken for our larger parts and our low volume smaller parts.

The second approach is used on smaller parts, parts that will fit into a 3 inch diameter, and still low volume users. Using this approach we punch all the holes for multiple parts on one sheet. Then rather than shearing the parts to separate them, we use a special punch & die to blank the part our of the sheet while still on the N/C punch. We make most of these special punches and dies on our wire feed EDM.

The third and fourth approach utilizes the N/C punches for the first run of the production parts, but subsequent orders for those parts are produced either using short run stamping dies or permanent 'hard' punch press tooling such as multi-station dies or progressive dies. The parts which would fall into this category would be high volume parts which have little chance of changing in the near future.

Figure 8

H.) COMPUTER AIDED INSPECTION

Because of the large number of machined castings and similar parts that are produced by our N/C machining centers, we must have a quick accurate method of inspection. We selected computerized coordinate measuring machines to solve this problem.

We operate these machines in a similar manner to our N/C machines. A quality engineer will decide how the parts should be checked. He will then prepare a setup sheet and write a computer program to check all of the holes and machined surfaces. The software in the computer is such that he can specify individual tolerances on each hole and machined surface and the computer will analyze them accordingly.

When the inspector starts to check a particular part, he loads the program into the computer via punched paper tape and sets the job up according to the setup sheet. He then follows a sequentially numbered drawing to run his checks. At each location or hole he inserts the probe and then steps on a foot pedal. The computer automatically takes the readings and analyzes them while the inspector continues the checking procedure. If the hole is within tolerance the computer will not print anything on the right hand side of the printout, if it is out of tolerance it will print which hole is out and the amount of variation. This printout along with the workpiece goes back to the machine operator so he can make the necessary corrections in his machine.

Productivity in our inspection department has increased as much as 800 percent using this system as compared to a standard coordinate measuring machine inspection of the same part.

Consistency is another big advantage that we have noticed from computerized coordinate measuring. The reports that the machine operator gets are always in the same format regardless of who inspects the part. The system eliminates the problem of each inspector interpreting the drawing in his own manner, infact the drawing simply becomes a reference document. When production parts are produced on our CNC machining centers and checked on our computerized coordinate measuring machines many times the drawing is not even used through the entire run.

Figure 9

I.)   COMPUTER AIDED PART PROGRAMMING

Because of our rapid growth and our large variety of N/C machines, we decided to use a timesharing computer system to produce our tapes. We decided on the system we have because the company that supplied it are the leaders in their field and because of the high quality software they supply. As the number of N/C machines increased we added to the programming system by purchasing an on-line graphics plotter. The graphics plotter software actually simulates the N/C machines motions on paper. This allows us to prove the programs out prior to running them on the machines. Using this method we minimize the machine down time and keep it in production a larger percent of the time.

As our tape library grew, storage of tapes and the ability to quickly edit them became more important. To accommodate this we installed a mini-computer with a dual drive floppy disk system. Each interchangeable disk, which is the size of a 45 rpm record, will store the equivalent of 2000 feet of tape. All of our master tapes are now stored on floppy disk. We continued to add to this system as our need increased and we presently have two high speed terminals with high speed reader-punches, one CRT display system and a graphic plotter.

We will continue expanding this system as long as we can see that it will help improve the productivity of our N/C machines. We are presently

considering interfacing our floppy disk system directly into our machine
controllers, thus eliminating the need for punched tape.

MANUFACTURING PROCESS PLAN

PART NUMBER	DESCRIPTION										
31108104	FOLLOWER CARD										
PROCESSED BY	PROGRAMMED BY	ECO NO.	REV	REV DATE	PREV TYPE	START DATE	DUE DATE				
H ILMER	HILMER			11-16-77	PROD						

OPER NO	OPER SEQ	TOOL NUMBER	WORK CENTER	LB GR	STANDARD HOURS MACHINE	LABOR	SET-UP HOURS	ITEM NO	COMPONENT PART NUMBER	QUANTITY		
10	010		SHR			500	2					
	020		SHR 4.15 X 7.00 (5052-H32 AL. ALY. SHT. .125 X 4 X 8)									
	030								70000800			
20	010	T1010P	PP			1 500	5					
	020		PIERCE & BLANK									
30	010		VM			2 000	5					
	020		MILL NOTCH (SEE SKETCH)									
40	010		PB			1 500	2					
	020		FORM COMPLETE									
50	010		TM			1 600	2					
	020		TAP (6) HOLES 4040									
60	010		VM			1 500	4					
	020		BORE (1) .50 DIA. HOLE									
70	010		FD			1 000						
	020		DEBURR .50 DIA. HOLE – DRESS TOP EDGE & BLEND TO RADII									
80	010		VIB			500						
	020		TUMBLE									
90	010		CF			1 000						
	020		CHROMATE 333P (IRIDITE)									
100	010		QC									
	020		INSPECT & ROUTE TO STORAGE									

Figure 10

J.)   COMPUTER GENERATED PROCESS PLANS

Once we realized the powerful editing capabilities of our minicomputer
system, as we applied it to N/C tape storage, we started to look for other
applications. The next logical application was our Manufacturing Process
Plans. This has allowed us much flexibility in updating and storage of
these documents and eliminated more of the space consuming file cabinets.

We visualize this data base that we are building as being an extremely powerful tool for future needs.

One of the uses for this data base will be a parts coding and
classification system. This system will identify parts by using an eight
digit number to describe the physical characteristics. This will allow
us to sort the parts by family, eliminate redundancy and optimize manufacturing processes. It will also allow us to sort process plans by
work center, giving us the ability to do better capacity planning.

SUMMARY

None of the individual systems alone can create a truly efficient manufacturing operation. It must be the seccessful integration of all the systems that are applicable to your manufacturing environment. It takes a management team that is engineering-oriented and willing to take the risks that are necessary to be the innovators of new processes.

The computer era is truly a part of the manufacturing scene, and it is not limited to the big companies with untold dollars to spend. It is available to even the smallest of job shops. The benefits that are to be derived are only limited by the creativity of the manufacturing management. Those companies that do not recognize this and use it to their advantage will not survive.

Effective CAM, DNC, CNC And NC System Maintenance

By Frank T. Cameron
Brown and Sharpe Manufacturing Company

Return-on-Investment is widely used to measure management performance particularly on large investments such as CAM, DNC, CNC and N/C system installations. In his new book, "Management Tasks-Responsibilities-Practices", Peter Drucker states as follows:

> "Effectiveness is the foundation of success - efficiency is the minimum condition for survival after success has been achieved. Efficiency is concerned with doing things right. Effectiveness is doing the right things."

An effective maintenance program efficiently operated is one of the cornerstones required to achieve your ROI objectives. Some of the key guidelines for effective maintenance include -

1. Utilization of available scheduled hours (Direct Workers) - 85% or more.

2. Allowable N/C System Downtime for Maintenance - 8% - 10% of available scheduled hours.

3. A good P/M Program. (Preventive Maintenance)

4. Good Lubrication and Filtration of Oils to 3 Microns.

5. Weekly reporting of performance.

The real cost of your CNC system downtime is the dollar value of shipments missed - probably $75/hour or more depending on the system installed. This perspective is essential to the development and implementation of an effective maintenance program for your N/C, CNC, DNC or CAM system.

In his recent book, "Management Tasks-Responsibilities-Practices", Peter Drucker states as follows (Chapter 4, page 45):

"_Effectiveness_ is the foundation for success - _efficiency_ is the _minimum_ condition for survival after success has been achieved. _Efficiency_ is concerned with doing things right. _Effectiveness_ is doing the right things."

(Underlining is mine)

Mr. Drucker's comments about _effectiveness_ and _efficiency_ go right to the heart of our maintenance challenge for any CAM, DNC, CNC or N/C system.

GUIDELINES FOR EFFICIENT MANUFACTURING PERFORMANCE

To build an effective maintenance program, you must first know the performance standards against which manufacturing performance is to be measured. Though each situation is unique, I should like to present the following parameters for your consideration. We shall then be on the same track for the balance of this presentation.

Performance vs. _Standard_ (Direct Workers) 95% - 100% using good measured day work standards.

Utilization of available scheduled hours (Direct Workers) _85% or more_.

Allowable Machine Downtime for Maintenance -

_6% - 8%_ of available scheduled hours for single purpose N/C machines.

_8% - 10%_ of available scheduled hours for machining centers and more complex systems.

(These downtime values include preventive maintenance, but exclude scheduled major machine overhauls.)

Please bear in mind with CAM, DNC and CNC systems, metalworking companies are rapidly moving from a _labor intensive_ to a _capital intensive_ situation. Capital generates ROI (Return-on-Investment) only while being utilized.

With this introduction let's consider some of the specifics of an effective maintenance program.

ELEMENTS OF AN EFFECTIVE MAINTENANCE PROGRAM

1. Management Awareness from Top to Bottom of the Need for Effective Maintenance

   (From CEO to Front-Line Supervision)

2. Know the Real Cost per Hour of Maintenance Downtime

3. Buy the Right CAM, DNC, CNC or N/C System for the Work to be Performed

4. Make a Proper Installation -

   (1) A good foundation.

   (2) Adequate and clean electrical service.

   (3) Adequate grounding.

5. Maintenance Organization Structure -

   Separate Planning from Doing.

6. Maintenance People -

   (1) Select the best.

   (2) Train them well.

   (3) Pay them well and keep them.
       (N/C technicians should be among the highest paid people in your shop)

7. Equipment and Supplies -

   (1) Your people need the right tools.

   (2) Establish a Maintenance Stockroom.

   (3) Provide adequate maintenance spares.

8. Provide a Maintenance Work Order System

9. Effective Lubrication and Regular Filtration of Oils is a Must!

10. <u>Preventive Maintenance Program</u> -

    (1) Develop one.

    (2) Implement it.

    (3) Modify it to achieve maintenance downtime objective.

11. <u>Report Performance Weekly</u> - Keep Score!

    (1) Ratio - Hours Run to Scheduled Hours.

    (2) Ratio - Maintenance Downtime Hours to Scheduled Hours.

    (3) Keep your people informed.

12. <u>Have a Periodic Equipment or System Rebuilding Program</u>

TOWARD AN EFFECTIVE MAINTENANCE PROGRAM

1. <u>Management Awareness from Top to Bottom of the Need for Effective Maintenance.</u>

Is maintenance just a "<u>necessary evil</u>" in the eyes of your Top Management?

Does your <u>Production Management</u> say, "We'll let you work on the machine or system - when and if we can spare it from production"?

Do operators say, "When the machine or system goes down, we'll holler <u>long and loud</u> until Maintenance gets it fixed"?

<u>Stop right here!</u> If this sounds all too familiar, your maintenance program is <u>most likely ineffective.</u>

The record will probably show -

(1) <u>Your maintenance downtime is high.</u>

(2) <u>Your productivity is below par.</u>

(3) <u>Your profit rate and ROI are below your objectives.</u>

190

To start changing attitudes, <u>try three places</u> -

(1) <u>Your tape readers</u> -

<u>Clean and service them weekly - a must!</u>

(Tape readers have been the source of much maintenance downtime)

(2) <u>Your tapes</u> -

Are they correct?

Maintenance people have spent much time only to discover a defective tape.

Try a test tape or on a CNC system check your input and output signals.

Have your programmer re-check his tape.

(3) <u>Filter your hydraulic oil to 3 microns each month</u> -

Dirty hydraulic oil causes endless wear and serious maintenance downtime.

Clean hydraulic oil will last 5 to 10 years or more without a change. Both Brown & Sharpe and the Los Angeles Division of Rockwell have proved this. <u>You can, too</u>!

The skeptical diehards will sit up and take notice as your maintenance downtime decreases and productivity improves.

2. <u>Know the Real Cost per Hour of Maintenance Downtime.</u>

<u>Your decision here is the key to the success or failure of your maintenance program</u>!

(1) Is it $9.00 per hour - the approximate cost of your direct labor plus fringe benefits for 1978?

(2) Is it $9.00 per hour plus factory overhead?

(3) Or is it the sales value per hour of goods not produced - say $60.00 per hour or more - maybe several hundred dollars per hour for a DNC system?

In considering the impact of maintenance downtime, take into account the following factors --

(1) The more heavily committed your shop is to N/C, CNC, DNC and/or CAM, the less opportunity you have to produce parts by conventional methods.

(2) Working overtime to make up for lost time is possible - if you are not already on an overtime schedule.

(3) Subcontracting of your production is possible - provided that other shops are not as busy as you are. Lead time is also a factor.

(4) If your equipment and work are unique, outside sources may be hard to find.

(5) If unplanned downtime results in a missed shipment and a lost customer, what does that cost you?

Should you conclude that your maintenance downtime cost is $100 per hour or more, your maintenance program will be far different than one based on $9.00 per hour.

This cost of maintenance downtime is a critical factor in building an effective maintenance program.

3. Buy the Right CAM, DNC, DNC or N/C System for the Work to be Performed

The correct system properly utilized means better productivity including better productive up-time.

(1) Give your maintenance people an opportunity to review the system you plan to buy.

(2) Locate other users and review their results with systems similar to the one you propose to buy.

(3) Consider the "Task Force" organization in purchasing and installing your new system. Include a maintenance person.

(4) Determine and evaluate the reliability in service of each major element and function of the system such as the control, tool changer, pallet shuttle positioning, machining functions, etc.

(5) Clarify your particular requirements, and include them in your original purchase order.

4. <u>Make a Proper Installation</u>

<u>Do it right</u> - <u>your troubles will be reduced!</u>

<u>Do it wrong</u> - <u>your troubles will multiply at an exponential rate!</u>

<u>Some Do's and Don'ts</u>

(1) Your foundation should weigh at least 1 1/2 times the weight of your machine.

Most machine tool builders are skimpy on foundation depths.

(2) Isolate your foundation from the factory floor with up to <u>2 inches</u> of styrofoam or other insulating material.

(3) Be aware of special vibration problems such as presses, railroads, heavy trucks, etc.

(4) Be sure your power supply to your machine and control are "clean". Don't connect to the same bus duct that services some welding equipment.

(5) Insure that you have a good ground for each control. Some builders require 5 ohms or less. Many PC Assemblies in controls cost $1,000 each. They are easily destroyed.

(6) Insure that your foundation is strong enough to keep the machine straight and level.

5. <u>Maintenance Organization Structure</u> -

Separate <u>Planning</u> from <u>Doing</u>.

Under a Maintenance Services Engineer, establish the following:

(1) A <u>Maintenance Work Order</u> System.

(2) A <u>Preventive Maintenance Program</u>.

(3) A <u>Lubrication and Oil Filtration Program</u>.

(4) A <u>Maintenance Storeroom</u>.

6. <u>Maintenance People</u> –

    <u>You can afford the best!</u>

    (1) Start with good supervision.

    (2) Select electronic technicians with good electronic backgrounds – the service technical schools, or electronic industries. (Your longterm maintenance electricians usually don't qualify.)

    (3) Your electronic technicians should be among the highest paid people in your shop.

    (4) Machine tool builders and control builders offer excellent training programs. If you have a number of people to train, consider having the instructors come to your plant to conduct the training. It's cheaper that way – more effective, too!

    (5) Let every technician – both electronic and mechanical – have his own manuals.

    (6) Provide good leadership to these skilled technicians.

    (7) <u>Help them to save steps.</u> Studies show that up to 50% of the time of maintenance people is spent walking.

7. <u>Equipment and Supplies</u>

    To be <u>effective</u> and <u>efficient</u>, good technicians need proper equipment and supplies.

    (1) Establish a maintenance storeroom and stock it adequately. Be guided by the real cost of downtime.

    (2) Use a maintenance clerk to establish and maintain your stock records.

    (3) You will need good oscilloscopes.

    (4) Each technician should have his own portable cart of tools.

    (5) Provide adequate alignment tools to verify system component alignments.

(6) Have manuals for each technician.

(7) Telephone the Service Manager of your machine tool vendor to ask questions and confirm your findings.

(8) Learn how to benefit from Emery Air Freight - VIP Service.

8. <u>Provide a Maintenance Work Order System</u>

To <u>schedule</u>, <u>dispatch</u> and <u>follow-up</u> maintenance work, a maintenance work order system is a must.

(1) Design your own form. There are many examples available.

(2) All maintenance work should be performed from a work order.

(3) <u>Schedule</u>, <u>dispatch</u> and <u>follow-up</u> on preventive maintenance work using the maintenance work order form. P/M work orders can be pre-typed for scheduled dispatching.

(4) Schedule and dispatch maintenance work using a Schedule-Dispatch Board.

9. <u>Effective Lubrication and Regular Filtration of Oils is a Must</u>!

Nobody needs to tell us how important good lubrication is. The real question is, "How do we insure achieving good lubrication?"

(1) With the help of your lubricant supplier, insure that you are using the proper lubricants.

(2) Establish daily "oil-routes". Oilers should be equipped with lubricant carts to enable them to lubricate machines on their routes.

(3) <u>Filter hydraulic and lubricating oils monthly</u> down <u>to 3 microns. Your oils should last for ten years</u>! Use your maintenance work order form to schedule this work.

(4) Clean oils and clean hydraulic systems will eliminate much downtime and replacement of expensive components.

10. A Preventive Maintenance Program

To achieve 85% productive up-time or better you must have an __effective preventive maintenance program.__

(1) Develop your own.

(2) Use the maintenance work order form.

(3) Schedule the required work weekly, monthly, quarterly, etc.

(4) Start your __P/M Program__ using information in the maintenance manuals from your system supplier.

(5) Your __P/M Program__ should include -

   a. Weekly or semi-weekly air filter changes.

   b. Lubrication and oil filtration to 3 microns.

   c. Weekly cleaning of tape readers.

   d. Checking all wipers.

   e. Clean electrical and electronic cabinets.

   f. Modify your P/M Program based on experience.

__Without a good P/M Program, your maintenance downtime will promptly double or triple!__

11. __Report Performance Weekly__ - Keep Score!

People like to know how they are doing. Two important measures are -

__Per Cent Productive Up-Time to Total Scheduled Hours__ (should be 85% or more)

__Per Cent Maintenance Downtime to Total Scheduled Hours__ (you can expect 6% - 10%)

(1) Publish the results weekly.

(2) Be sure your maintenance supervisors and technicians know what the results are.

(3) To achieve your ROI objectives you should equal or better the suggested targets.

(4) If you are consistently below target, find out why.

    a. <u>Correct the problems</u>.

    b. <u>Tighten up your P/M Program</u>.

    c. Your results should improve.

12. <u>Have a Periodic Equipment or System Rebuilding Program</u>

N/C systems work much harder than conventional machines - usually on a three shift basis. At the end of a period such as five years, your system will show signs of wear and deterioration.

Look for -

(1) Misalignment.

(2) Fishtailing of columns and saddles.

(3) Expended wipers - not wiping.

(4) Control circuit voltages near or below low limits.

(5) Dirt in control and electrical cabinets.

(6) Sluggish solenoid valves.

A thorough diagnostic check and major overhaul of this type will assure your continuing ability to meet your ROI objectives.

<u>CONCLUSION</u>

To compete effectively, you must -

1. Introduce and use efficiently the appropriate N/C, CNC, DNC or CAM system.

2. Achieve your stated ROI objectives.

An <u>Effective Maintenance Program</u> that is <u>Efficiently Operated</u> is essential to your success. Bon voyage.

APPENDIX

1. Procedure 1.1.3.12 - Filtration of Lubricating and Hydraulic Oils (pages 1 thru 3 and page 1 Enclosure A)

2. Mechanical and Electrical P/M Procedures and Frequencies for Warner & Swasey's 2-SC N/C Turning Machine

3. Maintenance Work Order - Procedure 1.1.3.8

4. Blank Weekly Performance Reporting Form

The <u>purpose</u> of this procedure is to establish the method and frequency for filtration of lubricating and hydraulic oils on designated critical production machines at Precision Park.

Equipment presently used for this filtration work is the IMP-356 (Imperial Hydraulics) equipped with a Brown & Sharpe pump and having a filtration capacity of 5 gallons per hour. This is a small, portable, electrically operated unit with three-stage filtration. These filters which are disposable are arranged in series as follows:

> 1st. stage - 100 microns
>
> 2nd. stage - 40 microns
>
> 3rd. stage - 10 microns
>
> 4th. stage - 3 microns

1. <u>N/C Machines</u> will be scheduled for monthly filtration of oil.

2. Other machines listed in Schedule A will be scheduled for bi-monthly filtration of oil. (Once every two months)

3. For each <u>5 gallons</u> of oil reservoir capacity, the filtration unit will be run for <u>one hour</u>.

The objective of this procedure is twofold:

1. By predetermined oil filtration schedules, keep lubricating and hydraulic oils clean and acid free and thereby extend the oil life to 5 years! (Past practice has been to change oil in critical machines yearly.)

2. By predetermined oil filtration schedules, keep lubricating and hydraulic oils clean and acid free -- thereby avoiding expensive breakdowns.

<u>Procedure and Frequency</u>

1. The Maintenance Services Engineer in cooperation with the Department Manager and the Maintenance Foreman will select and list critical production machines in which hydraulic and lubricating oils shall be filtered.

2. The Maintenance Services Engineer working jointly with Division Manufacturing Engineers and their designated representatives plus technical representatives from the oil vendor will determine the frequency of filtration.

3. The Maintenance Services Engineer will be responsible for scheduling and monitoring oil filtration work. Oil filtration work will be performed by machine repairmen assigned to such work by the Maintenance Foreman. At the beginning of each week, the Maintenance Services Engineer will issue to the Maintenance Foreman maintenance work orders (Form 821-1664) covering filtration of machines to be performed that week. Essential information required to complete the work will be pre-printed or pre-typed on the maintenance work order.

4. The Maintenance Foreman is responsible for maintenance of the IMP-356 oil filtration unit and maintaining an adequate supply of filter elements.

5. The Maintenance Services Engineer is responsible to establish and maintain an adequate file for completed work orders. Initially such completed work orders will be stored for a period not to exceed two years.

6. The Maintenance Services Engineer in cooperation with the Purchasing Department is responsible to arrange for chemical and physical analysis of between five and ten oil samples quarterly with the oil vendor. After the first year, at least half of the oil samples selected for analysis should be from the same production machines from which samples were taken a year ago. Written results of tests performed by the oil vendor should be reviewed with the Maintenance Manager and Division Managers of Manufacturing Engineering or their designated representatives.

7. Enclosure A is a list of critical machines at Precision Park on which oil filtration is to be performed.

8. Division Manufacturing and/or equipment Engineers are responsible for advising the Maintenance Services Engineer of additions or deletions to this list of machines requiring oil filtration.

9. The following list of lubricants are for general use in Brown & Sharpe Manufacturing Company:

Oils

Designation	ASLE Lubricant Type	B&S Stock No.	Vendor Description	Viscosity @ 100° Unless Specified Otherwise
B&S #8 Oil		840-48	Texaco	S.A.E. #30
B&S #51 Oil	S-32	840-59	Mobil Velocite #3	32 S.U.S.
B&S #52 Oil	S-60	840-60	Mobil Velocite #6	60 S.U.S.
B&S #53 Oil	S-105	840-61	Sunvis 701	100 S.U.S.
B&S #62 Oil	H-150	840-62	Sunvis 706	150 S.U.S.
B&S #63 Oil	W-150	840-63	Mobil Vacuoline 1405	150 S.U.S.

Oils

Designation	ASLE Lubricant Type	B&S Stock No.	Vendor Description	Viscosity @ 100° Unless Specified Otherwise
B&S #64 Oil	H-215 AW	840-126	Sunvis 747	200 S.U.S.
B&S #65 Oil	H-315	840-64	Sunvis 754	300 S.U.S.
B&S #71 Oil		940-65	Sunvis 7100	1500 S.U.S.
B&S #72 Oil	G-2150	840-66	Sunep 150	2150 S.U.S.
B&S #73 Oil		840-125	Texaco	S.A.E. 10W
B&S #81 Oil	W-315	840-67	Sunvis 500 Waylube #80	300 S.U.S.
B&S #82 Oil	W-1000	840-68	Sunvis #90	1000 S.U.S.
		840-133	Dexron H-36	

Review Responsibility: Manager of Maintenance

Distribution: All R. I. Holders of the Management Guide

Review Date: 6/1/79

NUMERICALLY CONTROLLED MACHINES

Dept. No.	Machine No.	Machine Description	Cap. in Gallons	Filtering Time(Hrs.)	B&S Oil Type	Oil Type	Location
Machine Tool Division							
5334	7-137	J&L Lathe	50	10	#62	Sunvis 706	P-14
5334	7-138	J&L Lathe	50	10	#62	Sunvis 706	Q-14
5334	7-139	J&L TNC Lathe	40	8	#62	Sunvis 706	Q-14
5334	7-518	W&S 2 SC	45	9	--	Dexron H-36	P-19
5334	7-519	W&S 2 SC	45	9	--	Dexron H-36	Q-19
5336	7-522	J&L TNC Lathe	40	8	#62	Sunvis 706	R-13
5335	10-178	Hydrotape	40	8	#62	Sunvis 706	N-22
5335	10-179	Hydrotape	40	8	#62	Sunvis 706	N-22
5335	10-236	VRM	50	10	#62	Sunvis 706	O-23
5335	13-64	5" Bullard	50	10	#64	Sunvis 747	M-25
5335	13-65	Sundstrand OM2	50	10	#62	Sunvis 706	O-23
5335	13-66	G&L	50	10	#62	Sunvis 706	P-26
5335	13-67	B&S Hydromaster	42	8	#53	Sunvis 701	N-23
5335	13-68	B&S Hydromaster	42	8	#53	Sunvis 701	N-23
5332	14-102	Behrens Press	200	40	#64	Sunvis 747	O-30

August 29, 1975

Code:
(7-1021)   5.29
(7-519)   5.30
(7-518)   5.31
(7-520)   5.33
(7-2006)   5.34

Preventive Maintenance Schedule
on Warner & Swasey 2-SC
Numerically Controlled Turret Lathes Model M-5040
with Mark Century 7542 Control
Machine No.'s 7-1021, 7-519, 7-518, 7-520 and 7-2006

The following maintenance intervals are suggested for use under optimum conditions. If environmental conditions warrant, the frequency of maintenance intervals should be increased.

Frequency	Service "Electrical"

Weekly

Monthly Air Filters: Air filters are located on the machine control cabinet. N/C control cabinet. Longitudinal servo drive and D.C. main drive motors. Inspect and clean the air filters with hot water and a detergent or any type of solvent and then blow them dry with air.

Monthly Control Panel:

1. Check for dirt, oil or water in control panel area and clean.

2. Check for airtight seal of control panel area.

3. Check tightness of screws on terminal boards and relays.

4. Check all switches and operating buttons including signal lights.

3 Month D.C. Main Motor
Servo Axis Motors
Index Drive Motors and D.C. Tachometers

1. Clean out any accumulated dirt and dust.

2. Check brushes.

3. Check brush spring tension.

Code: 5.29
5.30
5.31
5.33
5.34

Frequency	Service "Electrical"
3 Month (cont'd.)	4. Check commutator.
	5. Check mica between commutator bars.
	6. Check coupling tightness on Tachometer of feed-back devices.
3 Month	**Synchro Transmitters and Receivers:**
	1. Clean out any accumulated dirt and dust.
	2. Check brushes for excessive bouncing or arcing.
	3. Check excitation voltage.
6 Month	The flexible magnaloy coupling between the main drive motor and spindle drive transmission requires no lubrication, however, the rubber insert between the couplings should be inspected. Replace if deteriorated or worn.

Frequency	Service "Electrical"
Weekly	Lamp/lens: (Reeler only) Tape guides, reader photocell assembly, guide rollers. Dust, using a soft cloth, cotton swab, or brush. Clean using sparing amounts of pure isopropyl alcohol.
Monthly	Fixed/movable guide roller bearing: Sleeve type: Lubricate with one drop of an oxidation resistant mineral oil such as Gulf A&E or Toreso T-43 or T-52. Apply at the junction of the shaft and bearing. Caution: do not over-lubricate. Roller Type: No lubrication is required for the life of the components.
3 Month	Photocells in reel arm feedback assemblies: (Reeler only) Remove the cover over the photocell assemblies. Clean the faces of the photocells with a cotton swab dipped in pure isopropyl alcohol. Drive motor/reel motors. No lubrication is required for life of the components.
3 Month	Reel motor brushes: Check the length of them and replace them when their length is 5/16 or less. When the brushes are removed for checking, each brush must be replaced in the same holder from which it was removed.
Monthly	Reel shaft assembly: Check for excessive wear causing reel to be loose on shaft. Replace as necessary.
6 Month	Reader/Reeler: Check alignment and adjustments of the following: 1. Light line width and position. 2. Lamp voltage. 3. Solar cell output check. 4. Sprocket wheel.

Frequency	Service "Electrical"

6 Month (cont'd.)

5. Complete reading head alignment.
6. Tape tension arm spring.
7. Reel brake torque and wear check.
8. Photocell output and balance adjustment.
9. Reel motor phasing and direction of rotation check.
10. Motor zone switching points.
11. Broken tape lever switch actuation points.

6 Month

PWM-Servo System:

Re-tune the servo according to the procedure given in Chapter V.

6 Month

12 volt power supply.
Check and adjust the output.
Voltage should be within 2% of 12V.

1. Check for dirt or water in control panel area.

2. Check for airtight seal of control panel area.

3. Check tightness of screws on terminal boards and relays. (Use CAUTION)

4. Check all switches and operating buttons including signal lights.

Yearly

Resolver Feed Back Unit

Gear Train: Check, clean and lubricate the gear train with a light application of Mor Film Cling oil (250-300 SUS at 100°F).

Synchro: Inspect and replace if necessary.

Torque Motor: Inspect and replace if necessary.

Yearly

Control Cabinet "Cooling Surfaces":

1. Remove the side panels of the case and wipe the inside surface with a dry cloth to remove any dust or dirt accumulated by fans.

-5-

Frequency   Service "Electrical"

Yearly (cont'd.)

2. Reach into the air plenum first
    from one side of the control and
    then from the other, wiping the
    fins on top of the control until
    they are clean.

3. Clean the internal walls of the
    square tubes forming both side
    panels. Swab with a clean, dry
    cloth to remove accumulated dust
    and dirt. Since these surfaces
    are the least likely to cause
    trouble, this step is the final
    one to be taken.

V. S. Blair

August 29, 1975
Revised 11/7/75

Code:
(7-1021)   8.29
(7-519)   8.30
(7-518)   8.31
(7-520)   8.33
(7-2006   8.34

Preventive Maintenance Schedule
on Warner & Swasey 2-SC
Numerically Controlled Turret Lathes Model M-5040
Machine No.'s 7-1021, 7-519, 7-518, 7-520 and 7-2006

The following maintenance intervals are suggested for use under optimum conditions. If environmental conditions warrant, the frequency of maintenance intervals should be increased.

Frequency	Service "Mechanical"
Weekly	Check main hydraulic reservoir.
	Check slide way lubrication reservoir for level.
	Check cleanliness of ways and ball screws.
	Check oil level in air panel.
	Grease air cylinder packing gland.
	Grease chuck.
	Drain condensation from air panel.
	Check for leaks at hoses, tubings and fittings.
	Clean machine thoroughly and wipe down.
	Note: Do not use air for cleaning purposes. The pressure will drive dirt and chips into bearing surfaces.
Monthly	Check operation of mechanical interlocks.
	Check for grease leaks in motor and oil leaks from housing.
	Chip tote: Check oil level in reducer.
	Air control panel. Remove the filter, clean out the accumulated dirt and rust particles.

-2-

Frequency	Service "Mechanical"

3 Month - Millwright

Chip Tote: Conveyor
Inspect track and roller chain for
sign of wear. Make sure rollers turn
freely, lubricate if necessary.

Chip Tote: Clutch
Inspect clutch and adjust if necessary.

Chip Tote: Drive Unit
Inspect-make sure the "V" belts and
chain are aligned and slightly loose.

Chip Tote: Lubrication
Lubricate belt chain. Grease pillow
blocks. Change oil is speed reducer.
Oil drive chain.

6 Month

Main oil reservoir:
Change oil and filters.

Clean and wash all way wipers and
check wear condition. Replace if worn.

Main drive coupling. The flexible
magnaloy coupling between the main drive
and spindle drive transmission requires
no lubrication, however, the rubber
insert between the coupling should be
inspected. Replace if deteriorated or
worn.

Yearly

Replace way wipers as needed with
special attention to wipers adjacent
to metal cutting area.

Turret Index Drive. Check level of
grease in index drive transmission and
repack with fresh grease.

Gib and Caps:
There are three tapered gibs between
the bedways and the cross slide. One
under each way is held in place by a
bottom cap and a turret slide gib
between the cross slide and the rear
side of the right way. The gib setting
should be checked for correct adjustment as part of a regularly scheduled
preventive maintenance.

Code: 8.29
8.30
8.31
8.33
8.34

Frequency   Service "Mechanical"

Yearly (cont'd.)   Lubricate bearings in D.C. Main
                                   Drive Motor.

                                   Check level of grease in index drive
                                   transmission and repack with fresh
                                   grease per w/s lubrication specifica-
                                   tion.

V. S. Blair

Brown & Sharpe Manufacturing Company
AND SUBSIDIARIES

APPROVED BY:

PAGE 1 OF 2   PROCEDURE   1.1.3.8

SUBJECT

MAINTENANCE WORK ORDER

DATE ISSUED
4-1-68
DATE REVISED

Applies to: Precision Park Only   2-15-78

The <u>purpose</u> of this procedure is to provide an effective and economical means for planning, scheduling, and dispatching maintenance work. Maintenance Work Order Form No. 821-1664 will be used with this procedure.

This procedure will apply where the cost of maintenance work required does not exceed fifty (50) hours for labor or $500 for material. When either of these limits will be exceeded, the Work Order Form 457 will be initiated and approved per present practice.

Any foreman or department supervisor requiring maintenance work should initiate Maintenance Work Order Form No. 821-1664. The information to be entered by the originator is shown on the following sample:

B & S MAINTENANCE WORK ORDER FORM	TIME NUMBER OR EXPENSE ACCOUNT #		ORDER WRITTEN	
			TIME	DATE
	5334-317-316-12		10:30 a.m.	4-1-68
	DEPT. / SECTION	LOCATION	DESCRIPTION OF MAINTENANCE	
	EXPECTED COMPLETION DATE		REPAIR DEFECTIVE CLUTCH ON MAIN DRIVE.	
	4-3-68			
	ACTUAL COMPLETION DATE			
	TIME	DATE		
			SIGNATURE	
			N. E. SMITH	

FORM 821-1664

The originator will retain the original (white paper copy) and deliver the remaining two copies to the Maintenance Department.

Maintenance Work Orders involving installation of new equipment, relocation of existing equipment, and/or removal of old equipment will originate in the Plant Engineering Office. These Maintenance Work Orders will result from receipt of Machinery and Equipment Installation or Transfer Slips (Form 278) from the appropriate Division Director of Manufacturing Engineering or his designated representative. It is requested that the following lead time schedule be used in the issuance of these Transfer Slips:

1. Installation of standard equipment not involving replacement of existing equipment.   fifteen (15) days

2. Installation of standard equipment in-
    volving replacement of existing equip-
    ment.   thirty (30) days

3. Installation of equipment requiring
    special electrical or electronic
    controls.   Consult with Plant
                                                       Engineering allowing
                                                       lead time to purchase
                                                       long lead electrical
                                                       components.

4. Installation of equipment requiring
    a foundation.   ninety (90) days

Upon receipt of the Maintenance Work Order, the Maintenance Clerk will
stamp the date and time received on the second and third copies. The
appropriate Maintenance Section Foreman will check the expected completion date entered by the originator. Where the expected completion date
cannot be met, the appropriate Maintenance Section Foreman will contact
the originator to reach U & A on a different completion date. In a few
cases, U & A may have to be reached at a higher management level.

The estimated time prepared under the direction of the Maintenance Industrial Engineer will be used for maintenance work load determination by
trade.

The Maintenance Clerk will advise the originator of the order that the job
has been completed, and will mail the second copy to the originator showing the total actual time spent.

For maintenance work estimated to require less than 0.2 hours, the Foreman or Department Supervisors should call the Maintenance Department and
make necessary arrangements with the proper Maintenance Section Foreman.
For these few jobs, no Form No. 821-1664 will be prepared.

Review Responsibility: Manager of Maintenance

Distribution: Corporate, Machine Tool, PTG&P

Review Date: 6/1/79

N/C MACHS.	DOWNTIME REPORT	WEEK		WEEK		WEEK		WEEK		WEEK		WEEK		WEEK		WEEK		WEEK	
		HRS.	%	HRS.	%	HRS.	%	HRS.	%	HRS.	%	HRS.	%	HRS.	%	HRS.	%	HRS.	%
	Hrs. Scheduled																		
	Hrs. Run																		
	Maint. Down																		
	Prod. Ctl. Down																		
	Mfg. Engr. Down																		
	Inspect. Down																		
	Tools Down																		
	Dept. Down																		
	Hrs. Scheduled																		
	Hrs. Run																		
	Maint. Down																		
	Prod. Ctl. Down																		
	Mfg. Engr. Down																		
	Inspect. Down																		
	Tools Down																		
	Dept. Down																		
	Hrs. Scheduled																		
	Hrs. Run																		
	Maint. Down																		
	Prod. Ctl. Down																		
	Mfg. Engr. Down																		
	Inspect. Down																		
	Tools Down																		
	Dept. Down																		
	Hrs. Scheduled																		
	Hrs. Run																		
	Maint. Down																		
	Prod. Ctl. Down																		
	Mfg. Engr. Down																		
	Inspect. Down																		
	Tools Down																		
	Dept. Down																		

THE EXOTIC
NEW WORLD OF
DRIVES AND CONTROLS

Production equipment— macro improvements from microprocessors?

Courtesy, Kearney & Trecker Corp.

Those who view the microprocessor as the driving force behind a second computer revolution are predicting quantum leaps in machine control capabilities. Progress will be by evolution, counter the more conservative. Either way, production engineers will come out ahead.

By DONALD E. HEGLAND
Associate Editor

The use of industrial microcomputers will grow more than 20-fold in the next ten years, according to a recent study by Frost & Sullivan Inc. They further predict that today's biggest end-user of microcomputers—discrete parts manufacturing—will continue to embrace the technology while other segments of industry, notably process control and instrumentation, also become major users.

In terms of production equipment, IMTS '76 underscored this forecast. Numerical control permeated the show more than ever before with an unmistakable trend evident towards the highly sophisticated control systems. Over half the N/C units on display were CNC systems and the hottest trend in this area was clearly towards the use of microprocessors in control logic and programmable interfaces. Minicomputer and microcomputer-based CNC are fast assuming the dominant position on the numerical control scene and CNC is unquestionably where machine control is going. By 1980, it is expected that few, if any, machine or control builders will offer any other type of N/C.

At that time, numerical control, in the broad sense of the term—control by coded instructions, with no implications of tape, hardwire, softwire, etc.—will move into its rightful place in the manufacturing process and be accepted as *the* way to do almost any job, predicts Ex-Cell-O Corp.'s group vice president for machine tools, James L. Koontz. Four and five-axis contouring machines will become the standard machines of tomorrow, he feels, and hardly anything except numerically controlled machines will even be considered for use in manufacturing.

Another trend that will affect machine controls is the move away from centralized processing. Ac-

In five years, numerically controlled machines will be nearly the only type of machines even considered for use in manufacturing.

James L. Koontz, group vice president for machine tools, Ex-Cell-O Corp.

cording to National Semiconductor Corp., low-priced microprocessors will catalyze tremendous growth in multiprocessor distributed-intelligence systems. Distributed processing, using microcomputers, will be the wave of the future, according to the experts, and the world of machine controls will benefit from the spinoffs as distributed processing technology continues to mature.

Taking a different view, Kearney & Trecker Corp.'s manager of systems engineering and customer services, Richard Johnstone, feels that the microprocessor revolution is being somewhat overdramatized today. "The availability of microprocessors per se, those computers that are theoretically on a chip, is not the limiting factor in all the good things we're hearing about," he says, "it's primarily the software." The impact of the microcomputer will be evolutionary, he claims, because the software just won't come along that fast.

Where the microprocessor plus its dedicated software can be cranked out by the bushelbasket, as in microwave ovens for example, there may be a quantum jump. But so far as production equipment, specifically machine tools, are concerned progress will be evolutionary. "The dramatic changes are going to come," Dick Johnstone maintains, "because of the committed applications of intelligent LSI devices to manufacturing functions."

The small number of units involved is another deterrent to quantum leaps, according to Robert G. Chamberlain, vice president & general manager for electronics at Giddings & Lewis Electronics Co. He points out that there are only about 4,000 N/C and CNC units produced in this country each year, fragmented among several builders. The economics just aren't there in terms of developing a specialized product. "And," adds Dick Johnstone, "neither is that volume of business high enough to justify the requisite software effort."

Machines and controls will never be the same

Some authorities predict that machine controls designed around microprocessors will shrink physically and ultimately disappear, like Lewis Carroll's famed Cheshire Cat. Others see little, if any, changes in physical appearance compared with the controls we know today.

"The most visually obvious effect of building a machine control around microprocessors will be the demise of the floor-mounted control," argues Jim Koontz. Microprocessor-based controls are showing up on the columns of machines already, and if this trend continues, the result may well be machine controls physically invisible to the operator.

"We've gone through a glamour period," he adds, "when a few years ago, to sell a machine, it was necessary to make the control big, exciting, and mysterious. The

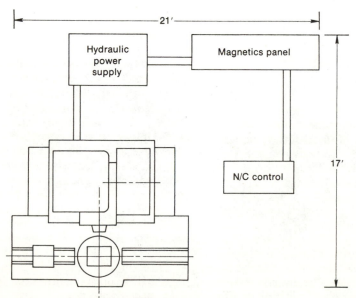

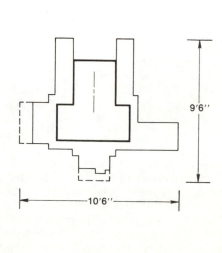

As the floor-mounted control shrinks and disappears into the machine, and as hydraulics give way to all-electric drives, tomorrow's machines will become simpler, more compact, and be thought of as just conventional machines, predicts Ex-Cell-O Corp.'s James Koontz.
Courtesy, Ex-Cell-O Corp.

trend today is the reverse—we want to see the control shrink literally to nothing. This is the direction I see it going and we're going to force it to go that way."

"I think we are simplifying the structural design of the machines dramatically," comments Dick Johnstone. "As a result of increased capabilities in the control system, we can reduce the numbers of gears, gearboxes, etc. by doing tricky things with the electronics and the software. That's where CNC has one of its biggest benefits. In fact, one of the biggest benefits of CNC is in simplifying the design of the ironwork."

Machines will take up less floor-space too, say the experts, and they will be easier to install. Two-to-four day startup will become fairly normal for a very sophisticated machining center, rather than several weeks or months. And these machines will no longer be the prima donnas of the shop floor—users will be able to relocate them easily.

The ability of microprocessors to compensate for inaccuracies in the machine will mean that things like machining centers will become more accurate and probably

cut into the precision machines, like jig borers, because machining centers will be capable of holding jig borer tolerances. In four or five years, people will be able to use the faster response times and better accuracies to push speeds and feeds higher. And machines will become more energy-efficient, partly as a result of using microprocessor-based controls. "The net result," predicts Jim Koontz, "should be that these machines will be less expensive. The user is going to get more machine for less money, for the first time in the machine tool industry."

Death knell for the tape reader?

Because 90% of today's problems with N/C relate to the tape reader, its elimination—by either storing part programs in PROMs or feeding them in from a mainframe computer—is an intriguing possibility. Opinions on the feasibility of this approach vary sharply, however.

"I think the day of tape will be

gone, and I think that within five years, tape and tape readers will no longer be necessary on the shop floor," says Jim Koontz. The small shops might use a portable tape reader, moving it from one machine to another to feed programs into the control memory. It's more likely, he feels, that they will use PROMs, preprogrammed on a mainframe and simply carried out and plugged into the microcomputer machine control. And the more sophisticated shops, he adds, will be strictly on mainframe computer via a direct link.

Parts programs on PROMs are not regarded so favorably by G & L Electronics Co.'s Bob Chamberlain. Such an approach, he feels, is analogous to the PC pegboards used before N/C, and is susceptible to the same logistics problems. Neither does he foresee the imminent demise of tape readers. "You're certainly going to put the program into storage and run from storage, as we currently are doing with this generation," he says, "and you could choose to come down from a host computer which would give you the possibility of removing the tape reader. But the people we've been talking to so

Microprocessor-based computer interface system by Beckman Instruments places process controllers under computer supervision with bidirectional digital data highway technique. Microcomputer in the controller assumes data acquisition and control functions automatically, without disrupting the process, when the host computer is off-line.
Courtesy, Process Instruments Div., Beckman Instruments Inc.

Distributed processing with multiple microcomputers in this numerical control system offers power and flexibility to serve different types of machine tools from two-axis lathes to five-axis machining centers. All software—machine logic, control programs, and control diagnostics—is stored in PROMs.
Courtesy, Actron Div., McDonnell Douglas Corp.

In terms of cost-effectiveness for the average contouring job, at this moment in time we just don't think that microprocessors are there yet.

Robert G. Chamberlain, vice president & general manager of electronics, Giddings & Lewis Electronics Co.

far—even though they want the capability of coming down from a host computer with a part program—want that tape reader left there so that if the host computer goes down, they'll still have a way of making their $50-100 an hour machine make parts."

"The marketplace is quite cautious now," adds G & L Electronics Co.'s senior electrical development engineer, Dr. Bruce Beadle. "We're not suggesting that some of these specific pieces of hardware will not come into being, but right now there isn't anyone beating down our doors to get direct computer interfaces and things like that. People are back-

ing themselves up. If they experiment with something like direct computer dispatch, they'll back it up with a tape reader until their own experience—and that of a lot of other people—tells them that the direct system is so reliable that they don't have to worry about it."

Can the micros really cut it yet?

Not in some areas. "We use a 32-bit minicomputer," explains Thomas B. Bullock, Giddings & Lewis Electronics Co.'s manager of electrical research & development, "because all the dimensions

on our machine tools are long enough that we can't put them in a single 16-bit word. If you were to do this same job with a microprocessor, you'd have to work in triple, maybe quadruple precision [i.e., you enter 32-bit words 8-bits at a time] because you don't have the word size, given the same architecture. So it's more economical to do it this way."

"Also, there are many more sequential operations involved in doing any calculation in multiple precision than in single," adds Dr. Bruce Beadle, "and the microprocessor is typically slower in execution speed, so the end result is less throughput."

Micros for various production machines

Use of microprocessor provides all the advantages of CNC at a fraction of the cost of minicomputer systems, according to Philips. Four-axis contouring numerical control can be programmed directly from the keyboard or by program transfer from magnetic tape cassette or punched paper tape.

Microprocessor control on Teledyne Pines bending machine holds bend angles within ±0.2 deg, distance between bends to ±0.005 in., and plane of bend to ±0.1 deg. Bending sequence and other data are stored in memory. Closed-loop circuitry provides positive positioning in bending mild steel tubing up to 3-in. OD by 0.019-in. wall.

Multiple-punch beamline with two-spindle drill station features multiaxis microprocessor control with multiple program storage that allows programming the next part while the machine is running the current part. W. A. Whitney machine has 100-ton punching capability in web and flange presses.

Courtesy, Teledyne Pines

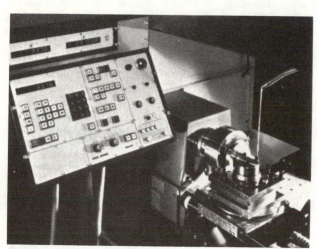

Courtesy, Philips

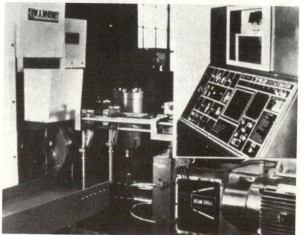

Courtesy, W. A. Whitney Corp.

❝ *The dramatic changes are going to come because of the committed applications of intelligent LSI devices to manufacturing functions.* ❞

Richard Johnstone, manager of systems engineering and customer services, Kearney & Trecker Corp.

What it adds up to, according to Tom Bullock, is that if you want to use a microprocessor where you're presently using a minicomputer, you can probably do it *if* the minicomputer is very under-utilized. But if your minicomputer is highly utilized, either time-wise or memory-wise, you're going to be in trouble.

Microprocessors are widely touted as being able to provide a less-expensive CNC unit. But Bob Chamberlain's opinion is that the cost is not significantly different—if it's different at all—so long as you're comparing the same functions. "We think that microprocessors are coming in CNC," he says, "but we think they're someplace down the pike. In terms of cost-effectiveness for the average contouring job at this moment in time, we just don't think that microcomputers are there yet."

"Fewer boards," is another advantage claimed for microprocessors. But that isn't always the case either. By the time you put in the things required to make the features identical, the discrepancy in the number of boards often disappears. As Bruce Beadle points out, the mini they use only occupies ten percent of the board nest. An equivalent system made up of microprocessors would probably occupy the same space, and have no impact on the other boards in the nest that handle input/output, interfacing, etc. "From our analysis," he says, "microprocessors don't give us any inherent advantage over the other hardware we're using."

"We think that peripherals are going to have much more of an impact on N/C in the next five years than microprocessors will," predicts Tom Bullock. "These are things like printers, punches, and terminals. We think you're going to see more use of these devices to connect management with the manufacturing floor."

Distributed control— some pros and cons

Distributed processing is the technique people are using to apply microprocessors to sophisticated control systems. One micro does the tape reading, another does the interpolation, and so on. Dick Johnstone is an advocate of distributed computing, which he sees as a growing philosophy. The advantage he cites is that you can have the ability to do bigger and better things with simple, intelligent LSI devices at the terminal level, whether it's a man-machine interface terminal in the classical sense, or a piece of a machine control for a machining center.

As an example, he notes their application of microprocessors to what they call an intelligent interface that will adapt virtually any machine tool peripheral device to the main system. And it accommodates new devices with only software changes.

Another advantage to having an intelligent terminal that can make decisions itself is that it takes some of the burden off the central system and frees it up to do other things. "There's no doubt," says Dick Johnstone, "that if there's one area where you're going to see the most universal adaptation of intelligent LSI devices, it's going to be in trussing up the old central computer; doing things that it's too busy to do."

Commenting on distributed processing within the machine control, Tom Bullock points out that functionalizing the control by splitting the operations up among several microprocessors changes the whole concept of the architecture and software of the control. "And we know," he goes on, "that in three years we'll have 16-bit microprocessors with ten times the speed of today's devices. When that happens, the central concept will be attractive again. So, in-

stead of causing confusion and extra training effort for our customers by changing the architecture back-and-forth, we'd rather wait the three years and then use a central microprocessor."

Functionalizing also creates a mini-network, with all the associated headaches. "And, although the idea of segregation sounds nice from an organizational point of view," cautions Bruce Beadle, "this forces each microprocessor to worry about a communications protocol with all the others—a function that doesn't exist in a single central computer."

Management information too

The intelligent terminal—a tool he can use to take the pulse of his production line while sitting in his office—will be the first place where the production engineer will get the "in-his-hands" benefit from microcomputers, in Dick Johnstone's opinion. Management feedback systems have always been thought of as being for sending production performance information to a management-type person for the classical managerial reasons.

In the future the production engineer is going to want to know, more than ever before, what his machine is really doing in terms of what he intended it to do. He'll become much more concerned with the uptime of the spindle on his high-cost machining centers than he was about the spindle on a single, stand-alone machine. We'll be seeing a different angle on management feedback, to a different level of people—the hands-on level—the production engineers. "You don't go into many factories today," he adds, "and see a production engineer with a full information system at his disposal, but we see that as the thing of the future."

Designing NC For High Productivity

By Thomas B. Bullock
Manager of Research and Development
Gidding and Lewis Electronics

ABSTRACT

Designing NC for high productivity situations demands that special consideration be given to reliability, ease of operation, ease of programming, and efficient use of computer time and memory.

This paper delves into the control designer's mind and explains his thought processes as he considers each of the above factors.

The presentation concludes with an overview of several challenging applications where many of the design considerations were especially important.

INTRODUCTION

If one is to achieve high productivity, it is imperative that the control have good reliability, that is, to be in operation a high percentage of the time. This translates into a high mean time between failures (MTBF) and a low mean time to fix (MTTF), both of which must be addressed early in the design. Operator's ease is the second most important consideration, since a control and machine will produce higher quality parts more rapidly when the operator feels comfortable and confident with the equipment. Productivity suffers when an operator must refer to a manual or puzzle over the method of entering information. Operation must be simple, straightforward, and easily understood. Improved programming, the third factor, will encourage the programmer/operator team to optimize the tape programs for better quality (i.e. better finish and accuracy) and minimum run time (by using shortest sequences and optimum feeds and speeds). Secondary benefits of improved programming techniques are the increase in programmer productivity and the decrease in the time required to initially verify tapes on the machine.

Careful planning of the architecture and instruction set of the central processing unit of the control allows more to be achieved in a given time and with a minimum amount of memory. In other words, the computer hardware is made more productive so that the return on the hardware investment is higher. Each of these four major categories will be discussed in more detail.

RELIABILITY

If a control is to be productive, it must be reliable. Reliability is defined as the uptime divided by the available time.

$$\text{Reliability} = \frac{\text{uptime}}{\text{uptime} + \text{downtime}} = \frac{\text{MTBF}}{\text{MTBF} + \text{MTTF}}$$

Historically, control designers have concentrated on MTBF (mean time between failures) during early stages of the design and have addressed MTTF (mean time to fix) as an afterthought. This has been a mistake as can be seen if we rewrite the reliability equation:

$$\text{Reliability} = \frac{\text{MTBF}}{\text{MTBF} + \text{MTTF}} = \frac{1}{1 + \dfrac{\text{MTTF}}{\text{MTBF}}}$$

It now becomes obvious that we can make the same improvement in reliability by reducing MTTF by a factor of 2 as we can by increasing MTBF by a factor of 2. In designing an NC for high productivity situations, both factors must be given significant consideration. As will be seen, reliability is the sum total of a large number of smaller design and test considerations.

First, consider MTBF and some of the factors which affect it. A primary consideration is strictly the number of components and backplane wires. A major benefit of CNC is an order of magnitude improvement in these two areas from the previous generation of controls. As an example, the Giddings & Lewis CNC control employs about a dozen boards and 280 backplane terminations for a four-axis control unit. The previous generation of controls took approximately 80 boards and about 5000 backplane terminations.

Another important factor is that circuits and control components be designed for high temperature operation. Experience has shown that a 168-hour dynamic burn-in at 125 degrees C. of all integrated circuits has reduced infant mortality problems by more than an order of magnitude. Core memory and tape readers are two other areas where temperature problems can exist. By specifying the core memory for 70 degrees C. operation and specifying conformal coating on the ferrite cores (to prevent deposit of cast iron dust), problems with this module can be minimized. Also, tape readers with 60 degrees C. temperature ratings are available as special units and should be specified in the design.

An alternative to specifying and designing all internal components for high temperature operation is the use of an air conditioner. This approach is not desirable from a designer's point of view because the air conditioner itself has proved to be unreliable. Also, when the temperature inside the control is below ambient, one encounters the problem of condensation inside the control on humid days. This condensation is detrimental in a number of ways.

A second problem with air conditioners is the constant cycling which stresses all components within the system. The technique of stressing components is a test tool for weeding out problems with new controls, but when done on a continual basis, components are eventually weakened.

Incoming inspection and heat cycling of major components such as CRTs, power supplies, and tape readers also helps ferret out weak components.

Final heat cycling of the entire control unit has a major impact on future reliability. My company has chosen to cycle all controls at their rated temperature limits for a minimum of 40 hours with the last 16 hours being trouble-free. There is nothing sacred about these numbers, but the experience is conclusive that simple spot-checking of controls in a hot box represents a compromise in reliability.

Many environmental reliability problems can be minimized by specifying a control enclosure which meets NEMA 12 standards. A dust-tight and drip-proof enclosure is much better in a harsh environment. Designing the operator's station to make it impossible to set a cup of coffee on it is a good practical suggestion.

Tolerance to conducted and radiated electrical interference are important design criteria. The NC unit must operate with electrical noise superimposed on the input power lines. IEEE Standard 472-1974 describes a high frequency and high voltage wave which any NC must be able to tolerate to consistently survive in an industrial environment. On the low end of the frequency spectrum, the Landis Tool Company noise generator has been accepted as a standard by most programmable controller builders; it also has proven to be a good test for NC equipment.

Shaking of sub-assemblies (such as backplanes) on a vibrating table during powered test of that sub-assembly has proven to be another technique for improving reliability.

Now, let's turn our attention to MTTF (mean time to fix). The most popular method for reducing the time required to bring a unit back on-line after a failure is diagnostics. Although input and output devices have historically been the most failure prone, the importance of MTTF suggests that one design a complete set of diagnostics. This set should cover the CPU, the memory, the CRT or other readout device, the tape reader, input/output, analog boards, the operator's station and the servo interface.

Also, extensive use of exerciser diagnostics should be considered for intermittent problems. For instance, rereading a tape and continuously verifying it against the same pattern in memory is an excellent check for intermittent tape reader problems.

Also to reduce MTTF, certain monitors and tests should be resident at all times so that access is immediate to certain basic information. At the same time, hardware costs should be considered in not having all items resident in view of the high reliability of CNC and the low probability of use. Examples of resident monitors are:

1. Motor overloads
2. Loop contactors
3. Hydraulic pressure switches
4. Oil pressure switches
5. Loss of feedback
6. Excess error
7. Over-travels
8. Command faults
9. Parity error
10. Servo command errors
11. Axes not referenced

When one of these faults occurs, it should be trapped in memory as it initiates a control shutdown. This way, when the control is restarted, one can determine what caused the shutdown in the first place. Of course, this requires that the memory be retentive during loss of power.

It is also desirable to have resident programs which allow one to monitor the status of all input and output devices, including pushbuttons, limit switches, triacs and feedback parameters. And the normal indicator lights for the operator's panel and messages to aid the operator should certainly be resident.

If possible, diagnostics should be structured to avoid disturbing the executive for as long as possible in troubleshooting and thus having to replace it. Generally, it is easier to replace part programs. Thus a hierarchy of diagnostics seems appropriate.

A good manual, a good set of diagnostics, and a spare parts kit can reduce the MTTF dramatically, providing a trained person is available to operate the diagnostics. Thus, a designer must provide data for good customer maintenance and programming schools.

In the event that trained personnel are not available, the ability to connect the numerical control to a service center by telephone to provide instant expertise is one alternative. An example is the system employed by Giddings & Lewis as shown in Figure 1. With such a system, one can monitor any devices connected to the control computer. He can also transmit diagnostics, run tests and retrieve results. We designed our system so that one portable terminal for connecting the control to the phone line is used per customer facility, in order to minimize the investment. This portable terminal

was designed to employ a microprocessor and thus service a
non-functional computer in the control. This service can
quickly pinpoint a problem to the board level, but, of course,
it is still necessary that the customer have a spare parts kit
available to make the desired replacements for minimum downtime.

OPERATOR EASE

A single operator's station with virtually all manual and
automatic controls near the point of action improves productivity by allowing an operator to run jobs faster and more
accurately. A joystick, easy manual data input (MDI),
incremental feed, feed and speed overrides, status indicators
and automatic tool set are all features which we feel should
be in such a station.

In addition, a CRT with all pertinent information allows the
operator to see those parameters which he wishes to monitor.
In designing the display of the CRT, one should select a
fixed format so that the operator knows exactly where to look
for a particular item. Information on modal dimensional
commands, actual machine locations, modal feedrates and speeds,
sequence number, active preparatory and miscellaneous codes,
and active actual tool offsets are all necessary information
to an operator, sometime during a day's operation.

Part program storage (PPS), which allows part programs to be
stored in the control itself, improves productivity because
the operator does less tape handling. In addition, a parity
error while reading will not have a disastrous effect (as
when a parity error occurs during a critical cut when running
directly from tape). Parts which are run repeatedly can be
stored indefinitely in the control. Once stored, the part
can be optimized to improve quality or save time. When
designing the PPS feature, it should be made expandable with
appropriate interfacing for future addition of DNC. Tape
bypass, to allow operating directly from tape when the PPS
area is full, is also very advantageous. The wear and tear
on tape and resultant problems when a parity error occurs
are very much minimized when part program storage is a standard
feature.

IMPROVED PROGRAMMING

Short and more accurate tapes which can be produced faster,
verified faster, and optimized easily contribute significantly
to productivity on the machine and in the programming department.

With contouring controls, both linear and circular interpolation should be available as standard. The feedrates
programmed should be the vector feedrates and not component
feedrates, simply because it is easier for both the programmer
and the operator. Being able to program full 360-degree

circles in a single block is also important to both individuals. Shop talk programming allows the operator to converse with the control in familiar terms such as RPM, IPM, IPR, and SFM (and the metric equivalents of these). This improves communication between programmer and operator, as they both understand the same terms. Further, it eliminates the code conversion tables used with earlier NC systems.

SFM (direct surface feet per minute) is very useful in lathe work. For maximum utilization of this feature, it should be designed to measure the radius of the cut to the tool tip. Otherwise significant discrepancies between true SFM and programmed SFM can result. The ability to incorporate IPR in conjunction with SFM also allows a consistent surface finsih when operating in this mode.

Designing inch/metric conversion so that it does not require re-referencing after switching can save considerable time for both the programmer and the operator. The ability to enter tool offsets in either mode and have them retained when modes are switched is also a timesaver.

Programming in both absolute (from a fixed part reference) and incremental (from a previous command) provides flexibility in allowing the programmer to utilize the dimensioning scheme which is simpler and faster for him.

Origin offset programming allows the programmer to temporarily shift his reference zero to make programming more convenient. As an example, when programming a bolt hole pattern, it is easiest to shift the reference temporarily to the center of the pattern. The result is simpler programming and easier understanding by the operator.

The use of tool offsets, cutter radius compensation and fixed cycles are other features to incorporate in the design which improve productivity of the machine and the programmer.

With computer-based technology in the control, the control designer can simplify the task of the programmer by encouraging him to identify similar operations on the part. The first time the programmer sees the operation, he identifies it as a sub-routine (or macro) and each time he requires it again, he simply calls the sub-routine. He only programs it completely the first time it occurs in the part. This has the added advantage in that all the dimensions in the sub-routine need be verified only once no matter how often the routine is used and transcription errors are reduced.

Similarly, where repetitious operations are identified, a loop may be set up to simply instruct the control to repeat the operation the required number of times.

With proper design, loops may be used in sub-routines and

sub-routines may be used in loops. A simple example of sub-routining and looping used together would be a threading pass. The pass would be defined with the thread lead, end points, retract, in-feed, etc.; if five passes are desired, the loop is instructed to repeat five times (in-feed would be in incremental mode). Such threading may have taken four instructions per pass previously (20 instructions in total) and now can be done completely in five instructions. If this is also defined as a sub-routine, it may be used over again later in the program with one single call instruction.

The use of these sub-routine, looping, and origin shift features will provide a great deal of "machining" with very short tapes. When these short tapes are stored in the program storage section of the control, little memory is used and more room is available for additional programs, or less memory needs to be purchased.

Threading routines, EIA/ASCII switchability and decimal point programming are other features to consider as standard in the design when productivity is important.

EFFICIENT USE OF COMPUTER TIME AND MEMORY

In some special applications, two of which will be mentioned later, it is imperative that careful consideration be given to computer time and computer memory. When all computer time is exhausted, the alternative is to add more hardware in the form of other controls or computers. Similarly, if memory is not carefully handled, the requirements of the job can expand into extra memory modules or, again, extra computers. In either event, the result will be a higher initial investment. In high productivity situations, we are not only concerned with high throughput, but also with return on investment, since rate of return can be considered the element which measures the productivity of investment.

Rather than list the attributes of a computer designed to make efficient use of computer time and memory, let's review the design that evolved in my company as we attempted to address these criteria. The need for considerable computational power which is relatively conservative of memory was recognized at the onset of our design. As a result, a special purpose computer was designed for machine tool, process control and special machine applications. The basic structure of this computer is shown in Figure 2.

The computer consists mainly of three fundamental parts. The first part is an extended arithmetic capability unit (a 32-bit computational mathematics section). Since most dimensions and computations encountered in machine tool applications will not fit into the standard 16-bit word, those using 16-bit general-purpose computational units must consistently work with double precision. A simple, double-precision addition with a standard, general-purpose mini-computer can take several computer cycles to do what this 32-bit design can do in one computer cycle.

The second element of our processor is the 1-bit logic capability. This unit cooperates with a portion of the mathematical unit, but accumulates results independently. Separate steps of chained Boolean operations may be done in sequence or may be interrupted and continued later. When logic is accomplished with a general-purpose mini-computer, it uses the same accumulator and circuitry to perform that logic. The result is that a general-purpose computer must make either a mathematical or logical computation and store that result before it can switch to the other type of computation.

With the design being described here, a mathematical computation can occur and the answer held in the math accumulator, it is then possible to do logic which will result in a logic bit being stored in the logic accumulator.

The third major component in this design is the status word. The status word is a word of memory which allows the two accumulators to talk to each other. It permits mathematical calculations to be conditional upon logic results and also to make logic statements conditional upon results of mathematical computations. As an example of this, if the accumulator is greater than the input (AGB), bit 0 of the status word will be set. Referring to Figure 2, it can be seen that bit 0 reflects this condition. Bit 1 reflects the condition when the accumulator is equal to the input (AEB). Similarly, other mathematical and logic conditions can be stored in the status word.

Since most machine control requires considerable inter-relationship between the mathematical and logical worlds, this design becomes ideal for process and machinery control.

In order to conserve memory, the design was structured so that each logical word of memory would contain 16 bits of logical information. The memory reference instructions can now access any of those 16 bits in any one of 64 words of memory devoted to this storage as shown in Figure 3:

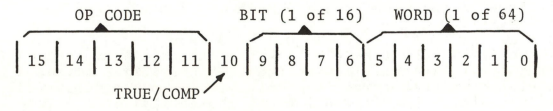

Figure 3

As can be seen, the five most significant bits are devoted
to the operation code for such things as LOAD, AND, OR, SET,
SAVE, ETC. The number 10 bit indicates whether the operation
will occur with the true or complement of the bit selected.
Bits 6 through 9 of the memory reference instruction single
out the bit within the word. The word is selected with bits
0 through 5 for a possible 64 different words (resulting in
1024 bits or flags). The logic for selecting the correct
bit within the word and eventually re-writing that bit is
shown in the logic portion of Figure 2.

The net result of the above is that through careful design,
a powerful computational and logical capability is possible
with a reasonable amount of memory and a single central
processing unit (for greater reliability).

APPLICATIONS

The Giddings & Lewis CNC 800 control which employs the GL3200
computer mentioned above is used for all of our standard
2-, 3-, and 4-axis machines for turning, milling, drilling,
and boring. The thought processes described earlier under
reliability, operator ease, and improved programming
techniques have been employed in that control.

In addition to these more standard applications, special
applications have arisen which are of particular interest.

One of these special applications is a control used in the
corrugated box industry. This control operates a 42-axis
flexo-folder/gluer. This is a machine which takes a stack of
corrugated material cut to some standard size, and trims it,
slots it, creases it in the appropriate places, folds it,
and glues it to provide a stack of flattened boxes which
are ready to be sent to the final user.

Interestingly enough, the control was justified strictly from
the improvement in setting up to run boxes. The following
chart (Figure 4) compares a conventional box making machine
to one with a CNC unit on a two-shift or 16-hour basis. It
is assumed that the productive time during that period is 14.4
hours. As can be seen, the CNC unit reduced the unit setup
time from .6 hours to .2 hours, thereby allowing 14 box runs
rather than 10 during the normal two shifts. The net result
is box production of 104,400 units instead of 75,600 units
as normally achieved. This represents a 38% increase in
productivity.

Another interesting factor in this application is that only
8 pieces of information were required to totally define a
box. The central processing unit takes these 8 dimensions
and calculates the required moves of all 42 axes. As one
might expect, great computational power and a good inter-
relationship between logic and mathematical worlds was

necessary for the successful completion of this application.

Another application involved ten axes of simultaneous contouring. The machine is a two-headed welding seam tracker.
Each head has three linear motions and two circular motions.
The purpose of the circular motions is to allow the welding
wire to be positioned perpendicular to the seam at all times.
The two heads would be simultaneously welding similar seams.
In this application, the servo loops were closed in software
with each of the ten axes being upgraded every eight milli-
seconds.

The key to achieving a high return on a minimum investment in
these special applications rests with a special-purpose
computational and logical unit designed specifically for
machine and process controls.

CONCLUSIONS

Reliability is a primary design factor for equipment which
is to be used in high productivity applications. Good
reliability means a high MTBF and a low MTTF. Operator
ease and programming advantages improve both quality and
productivity. Efficient use of computer time and memory make
difficult applications possible with a single control and
computer to keep the investment low.

NumeriEAsy Teleservice (NEAT)[T.M.] diagnostic service

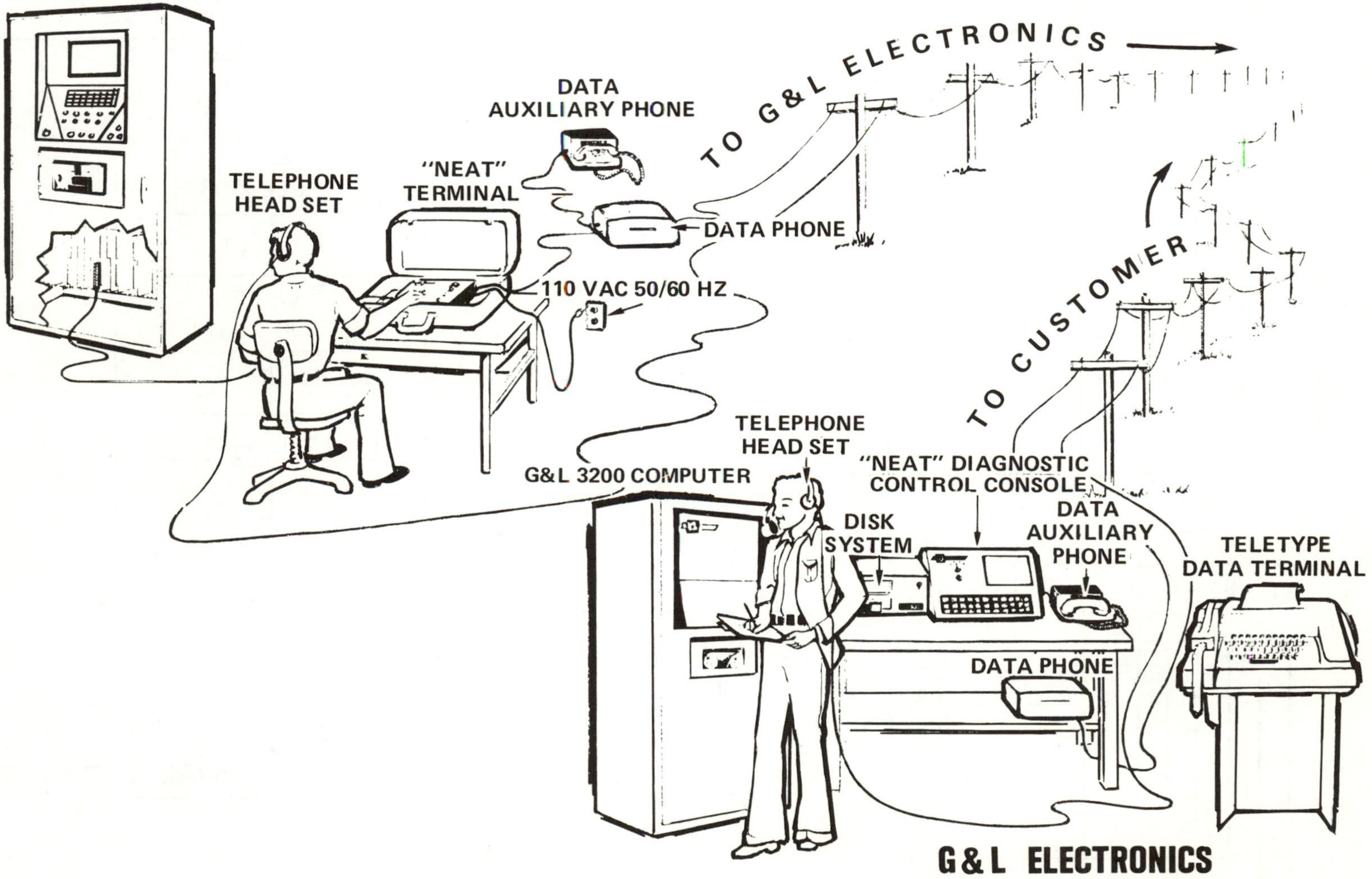

CNC CONTROL

TELEPHONE HEAD SET

"NEAT" TERMINAL

DATA AUXILIARY PHONE

DATA PHONE

110 VAC 50/60 HZ

G&L 3200 COMPUTER

TELEPHONE HEAD SET

TO G&L ELECTRONICS

TO CUSTOMER

"NEAT" DIAGNOSTIC CONTROL CONSOLE

DATA AUXILIARY PHONE

TELETYPE DATA TERMINAL

DISK SYSTEM

DATA PHONE

G&L ELECTRONICS

Figure 1

229

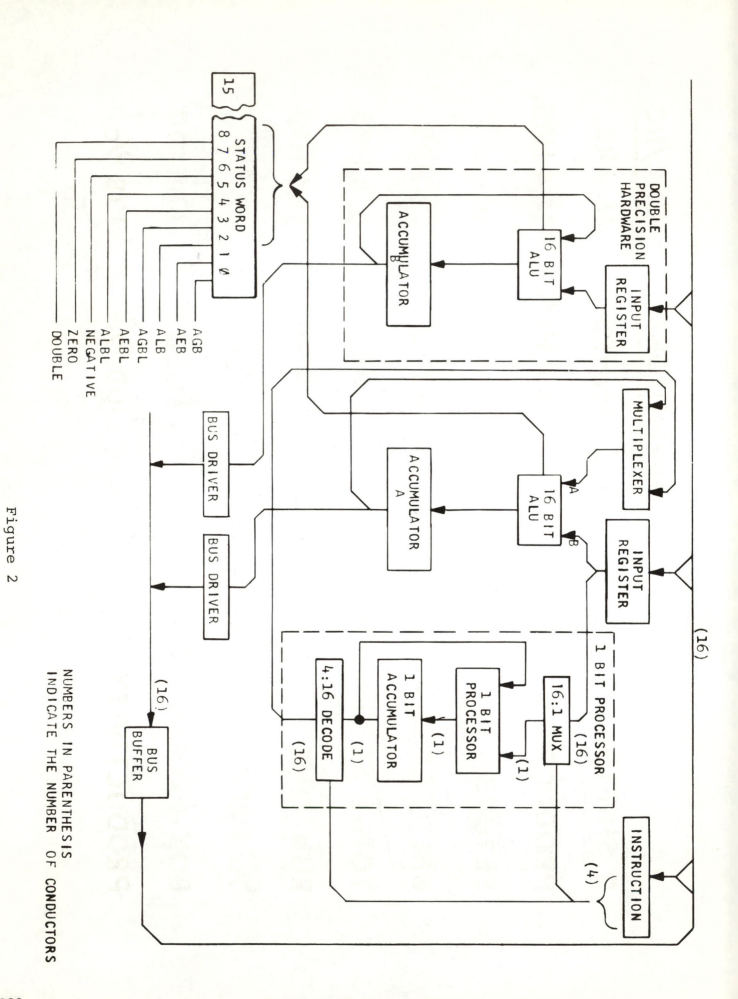

Figure 2

NUMBERS IN PARENTHESIS
INDICATE THE NUMBER OF CONDUCTORS

230

	CONVENTIONAL	CNC/NC
2 SHIFTS	16 HRS.	16 HRS.
PRODUCTION TIME	14.4 HRS.	14.4 HRS.
SETUPS	10	14
UNIT SETUP TIME	.6 HRS.	.2 HRS.
TOTAL SETUP TIME	6 HRS.	2.8 HRS.
RUN TIME	8.4 HRS.	11.6 HRS
% PROD.TIME FOR SETUP	41 %	19 %
BOX PRODUCTION	75,600	104,400
PRODUCTIVITY	100 %	138 %

Figure 3

231

CHAPTER 6

MAINTENANCE

Reprinted from: Modern Machine Shop, February 1978

A Direct Numerical Control System Reduces Downtime

A direct numerical control system is being installed on a real-time basis to remove the environmental factor from NC machine tool operations. Initial results indicate substantial savings and increased productivity will be obtained.

By ROBERT GAMRATH, Director of Facilities, and
FRED STEWART, Manager of Mechanical Equipment Engineering
Fabrication Division
Boeing Commercial Airplane Company
Seattle, Washington

Numerically controlled machine tools are something of an enigma. They are designed to substitute automated, precision cutting for the human frailties that often inject themselves into manual machine tool operations. Yet, the harsh environment of the machine shop they operate in often causes failures that tend to dull the NC concept.

Because the Fabrication Division of Boeing Commercial Airplane Company, Seattle, Washington, can do little about the machine shop environment, a major cause of downtime—tape readers and spoolers—are being removed from the shop floor. They are being replaced with a minicomputer-based, on-line, direct numerical control system. It is planned to have 80 machine tools assimilated into the system by the end of 1979.

At present, 25 machine tools are being operated with the direct numerical control (DNC) system. Enclosed in a room above the machine shop floor, the computers are removed from the severe elements and, therefore, are unaffected by the environment below.

Although the system is still many months away from full implementation, its present phase is far and away superior to past methods. Like most machine shops with numerically controlled tools, some photoelectric tape readers and electromechanical spoolers are still located on the shop floor.

The Mylar program tapes, which range from 60 to 1200 feet in length, were previously brought to the shop floor where they were run through readers. The average job required 600 feet of tape. Programs were read in blocks, with the tapes starting and stopping after every one to two inches. Thus, in a 500-foot reel of tape operating at 400 characters per second, the reader started and stopped 12,000 times.

Computer operator installs punched tape onto the tape reader of an Allen-Bradley 1795 series computer system. Heart of the system, which controls the metal-working tools in the machine shop, is an Interdata 7/32 minicomputer.

Besides such heavy-duty operation, the readers and spoolers were exposed to hydraulic and lubricating oil, coolant, metal chips and a mist that hangs over the shop. All of this contributed to numerous failures and an ongoing program of maintenance and replacement.

As conceived in diagram form, the new DNC system will resemble a pyramid with part-programming computers at the apex servicing two modules that will operate a total of 80 machine tools.

A module consists of two Bulletin 1795 series computer systems, developed by Allen-Bradley Company, Cleveland, Ohio. Each system has its own Interdata 7/32 minicomputer that transmits instructions from a remote location directly to the machine tools. Each 1795 series computer is designed to operate 20 machine tools. The computers may be linked to machine control units, which in turn operate the machine, or to machine terminal units that are connected to machine control units. The

Programs are called from an active library by machine tool operator at console connected to the 1795 series computer system. The system has reduced downtime caused by failures of punched tape readers and spoolers on the machine shop floor.

Machines ranging from this massive cutting tool to small lathes are incorporated into the direct numerical control system.

isolated computers are hardwired to the machine terminal and machine control units on the floor. The nearest machine is 40 feet away, while the farthest is 200 feet from the computer room. A hardwired connection can be made up to a distance of 5000 feet.

When the system is complete, parts programs will rest in a data base library, also resembling a pyramid. At the apex will be an "active" library of stored workpiece programs, which can be quickly accessed for transmission to a tool-operating terminal.

The "ready" library in the middle will be for workpiece programs scheduled for near-term use, but not currently in demand by a terminal. Speed is less important for the ready library, but still must be at a suitable rate for instant response to terminal operators.

The "archives" are exactly that—every workpiece program in the library. Most would not be scheduled for use and, therefore, rapid accessibility would not be a factor.

At present, only the active and ready libraries are maintained. Each ready library has the capacity for 50,000 feet of punched tape. The active library can hold data equivalent to 24,000 feet of punched tape. The archives will not become a part of the system until mid-1979.

When the system is totally operational, the manufacturing engineering group responsible for developing workpiece programs will store the data in the archives. The workpiece programmers will have sole responsibility for management information in the archives.

Programmers will be able to revise the status of programs or purge them from the archives. Using a video display terminal, a programmer will be able to track a workpiece program to a particular library level and put an immediate hold on it for change or purge. If the program is already in the active stage, machining of the workpiece will be completed before the program is recalled.

By the end of 1978, station terminals will be located on the shop floor.

At that time, job kitters (the person on the shop floor responsible for gathering manufacturing data, documentation, Mylar tapes, instructions, and so on, in kit form) will insert punched cards into the terminals to call workpiece programs from the archives to the ready stage. A two-hour span has been allocated to move a workpiece program from the archives to ready. That's the average time it takes to set up the machine tool and bring the workpiece to the floor.

In the interim, the operator gets his information from documents supplied by the job kitter. When the operator is ready to machine the workpiece, he enters a request through the keyboard of his console and the program is moved from ready to active. On completion of the job, the operator, through his console, purges the program from active and the kitter, through his terminal, purges it from ready. Obviously, the intent is to completely automate all machine tool operations. **MMS**

Reprinted from: Machine and Tool Blue Book, August 1979

Up Your Machine Time With Preventive NC Maintenance

Planned maintenance program keeps the lid on the potentially high downtime costs of advanced manufacturing equipment.

By BERNHARD BARON, VP & Plant Manager, DSM Manufacturing Co., Denver, Colo.

A reduction in downtime on its key 20-station NC punching machine has improved overall production efficiency an estimated 50 percent at DSM Manufacturing Co., Denver, Colo. With the adoption of highly automated equipment and techniques, the high volume producer of precision electronic chassis and control panels has experienced accelerated growth and a tripling of sales volume.

Cornerstone of the increased production capability is the NC turret press. Although the unit was installed to increase capacity and not replace existing machines, the unit produces more than five manual machines. In addition, the machine's ability to accurately punch panels, 20 at-a-time, has contributed significantly to the reduction of scrap.

Balancing Act

Because the NC machine is so highly productive, it cannot be run full time. However, the shop is geared to a production level that could not be manually achieved, in practical terms, should the NC unit become inoperable. A major breakdown would inevitably translate into work flow interruptions, revised scheduling and late deliveries.

This domino effect would also require suspending a number of vital operations such as shearing and punching. Then, once the machine is repaired, all of that preprocessing work would have to be made up. This, in turn, supplants the time normally allocated to other production operations such as deburring and forming. Thus the production slack created by a one or two-day breakdown often takes a month to overcome.

Preventive Maintenance

An evaluation of DSM's particular requirements led to a contract with Honeywell's Process Control Division, Ft. Washington, Va., which provides a nationwide NC maintenance service. The program includes both preventive and corrective maintenance on all electronics and low-level magnetics (relays and interfaces between the controller and the machine).

The initial advantage was the availability of immediate service from a local service office (the network includes over 100 offices throughout 42 states). A breakdown, in all probability, can now be resolved in terms of hours rather than days ... should one occur. But the long-range benefit of the plan was the institution of a practical preventive maintenance (PM) program.

For example, during their first check Honeywell technicians concentrated on restoring the machine to its original design capacity, repairing and replacing worn and weakened components and making the necessary consecutive adjustments to fine tune the unit to peak precision. Since then it's been largely a matter of PM follow-through which, in itself, is no small operation. The checklist is extensive (note sidebox).

Electronic chips and circuit boards are checked frequently as are heat levels and variations in heat buildup. Sometimes only a small variation in heat buildup can alter motion and distance commands to the machine. Minor variations in the performance of capacitors and resistors can materially affect the close tolerances inherent in the system.

Service technicians are always looking for potential trouble-spots. If they recognize that a capacitor is becoming weak, for example, it is changed on the spot rather than waiting, say, six weeks until it fails and a half day in production is lost. The essence of such main-

Preventive NC Maintenance

tenance is the prevention of unexpected downtime. A certain amount of

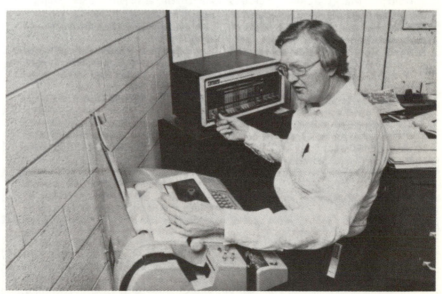

NC programming efficiency depends upon regular maintenance checks so that punched paper tape output is provided with the accuracy that is required for new setups and tooling changes. A Honeywell technician checks the electronic circuitry of this equipment located in the offices of DSM Manufacturing Co.

A 40 percent increase in machine availability has been achieved as the result of a regular preventive maintenance program on a key, 20-station NC turret press used in the production of precision electronic chassis and control panels.

NC SERVICE PROGRAM

Scope: Program provides complete service on machine control units, operator stations, servo drives and tape preparation and editing units, regardless of make. Service is comprised of (1) routine maintenance, (2) corrective maintenance and (3) replacement parts and spare parts inventory. In addition, an operator log is provided for each NC machine.

Routine Maintenance: Includes minor and major servicing, scheduled at planned intervals, to meet the needs of each installation. Minor service, performed by the service technician includes:
- Air Filter—Inspect, clean or replace.
- Console—Inspect, clean and check all control panel operations and replace faulty switches and indicators.
- Correct any minor problems noted in the operator log.

Major service is performed less frequently, but is more comprehensive and includes:
- Check and adjust all power supply levels.
- Check and adjust servo offset balance, overlap and power.
- Check and clean circuit-board sockets, plugs and contacts.
- Check, clean switches.
- Check and adjust feed rates.
- Check and adjust tachometer feedback.
- Check and adjust dwell timers.
- Check and adjust spindle speeds.

- Check data input device.
- Check and adjust turret position electrically.
- Clean and replace filters.
- Check and adjust resolver power supplies.
- Check and adjust the feed rates of all axes to the same level.
- Check and set the arc center coordinate registers.
- Replace or repair any marginal components as required to insure proper operation.
- Check master clock frequency and wave shape.
- Inspect, clean and check tape reader including bulbs, mechanical fingers, capstan, take-up and feed spools, speed drives, and tape-reader levels.

Corrective Maintenance: Normally provided within 24 hours after notification of machine failure. Service includes complete system diagnostics for electronic, mechanical and hydraulic failures.

Replace Parts/Inventory: Service vendor supplies all electronic replacement parts. Spare parts inventory includes printed circuit boards, tape reader assemblies and parts, feedback devices, timers, limit switches, low-level magnetics, lights, registers, and any other parts required to expedite repairs. Parts inventory is maintained either at the user plant or in service company's local, branch or central parts depot.

To obtain complete details on this Numerical Control service, **circle No. 7 on** reader service card.

---

downtime—for corrective maintenance, tool sharpening and mechanical refurbishing—is neither avoidable nor debilitating if it's planned. But emergency breakdowns always happen at the wrong time—usually in the middle of an important job or when rushing to meet a critical delivery date.

Production Payoff

Since the program began over two years ago DSM has not experienced a single major shutdown that materially affected production. There have been breakdowns, however, which required emergency maintenance and the machine was frequently shut down at the outset for corrective maintenance. But downtime was counted in hours instead of days. Then as production employees became more accustomed to working with the servicemen they learned more about the machine's control capability. On a number of occasions—even if the machines were running reasonably well—a technician would be called for additional adjustments to enhance performance efficiency.

Prior to instituting the present service and maintenance program, DSM could realistically project the NC machine's operation at approximately 50 percent of its potential availability. However, that figure was inconsistent. Some days it would run at 70 percent, other days at 20 or 30 percent. And depending on the work flow at that time, the reduced levels of operation could and did cause production bottlenecks.

Machine availability has greatly improved and is now operated at a consistent 85 percent—quite satisfactory in the face of mandatory downtime for tool changes and new setups in addition to maintenance requirements. But perhaps more important, is that the 50 percent increase in machine availability is at least duplicated in overall plant productivity and efficiency. • • •

INDEX